Universitext

Universitext

Universitext is a series of textbooks that presents material from a wide variety of mathematical disciplines at master's level and beyond. The books, often well class-tested by their author, may have an informal, personal, even experimental approach to their subject matter. Some of the most successful and established books in the series have evolved through several editions, always following the evolution of teaching curricula, into very polished texts.

Thus as research topics trickle down into graduate-level teaching, first textbooks written for new, cutting-edge courses may make their way into *Universitext*.

For further volumes:
www.springer.com/series/223

Luis Barreira · Claudia Valls

Dynamical Systems

An Introduction

Luis Barreira
Departamento de Matemática
Instituto Superior Técnico
Lisbon, Portugal

Claudia Valls
Departamento de Matemática
Instituto Superior Técnico
Lisbon, Portugal

Translation from the Portuguese language edition:
Sistemas Dinâmicos: Uma Introdução by Luís Barreira and Claudia Valls
Copyright © IST Press 2012, Instituto Superior Técnico
All Rights Reserved

ISSN 0172-5939   ISSN 2191-6675 (electronic)
Universitext
ISBN 978-1-4471-4834-0   ISBN 978-1-4471-4835-7 (eBook)
DOI 10.1007/978-1-4471-4835-7
Springer London Heidelberg New York Dordrecht

Library of Congress Control Number: 2012953900

Mathematics Subject Classification: 37Axx, 37Bxx, 37Cxx, 37Dxx, 37Exx

Printed on acid-free paper

Springer is part of Springer Science+Business Media (www.springer.com)

Preface

This book is a self-contained introduction to the core of the theory of dynamical systems, with emphasis on the study of maps. This includes topological, low-dimensional, hyperbolic and symbolic dynamics, as well as a brief introduction to ergodic theory. It can be used primarily as a textbook for a one-semester or two-semesters course on dynamical systems at the advanced undergraduate or beginning graduate levels. It can also be used for independent study and as a rigorous starting point for the study of more advanced topics.

The exposition is direct and rigorous. In particular, all the results formulated in the book are proven. We also tried to make each proof as simple as possible. Sometimes, this required a careful preparation or the restriction to appropriate classes of dynamical systems, which is fully justified in a first introduction. The text also includes many examples that illustrate in detail the new concepts and results, as well as 140 exercises, with different levels of difficulty.

The theory of dynamical systems is very broad and is extremely active in terms of research. It also depends substantially on most of the main areas of mathematics. So, in order to give a sufficiently broad view, but still self-contained and with a controlled size, it was necessary to make a selection of the material. In view of the necessary details or the need for results from other areas, some topics have been omitted, most notably Hamiltonian and holomorphic dynamics, although we have indicated references for these and other topics. We have also provided references for further reading on topics that are natural continuations of the material in the book. These suggestions, together with a short description of the contents and of the interdependence of the various chapters, are grouped together in the introduction.

Lisbon, Portugal   Luis Barreira
                                                           Claudia Valls

Contents

Chapter 1
Introduction

This introductory chapter is a user's guide for the book. It includes a brief description of the contents of each chapter and of the interdependence of the various topics. It also includes suggestions for further reading and for specific courses based on the book.

1.1 Contents and Suggestions for Further Reading

We first summarize the contents of each chapter. We emphasize that we do not strive for completeness. Instead, the idea is to give a brief overview of the contents, highlighting the main topics and pointing out those that can be considered more advanced. We also indicate references for further reading on topics that are natural continuations of the material in the book.

1.1.1 Basic Notions and Examples

Chapter 2 forms the basis for the rest of the book. It is here that we introduce the notion of a dynamical system, both for discrete and continuous time. We also describe many examples that, together with other examples introduced throughout the book, are used to illustrate new concepts and results. The examples include rotations and expanding maps of the circle, endomorphisms and automorphisms of the torus, and autonomous differential equations and their flows. In addition, we describe some basic constructions that determine new dynamical systems. This includes going back and forth between discrete and continuous time.

The emphasis of the book is primarily on dynamical systems with discrete time, although we still develop to a reasonable extent the corresponding theory for flows. A natural playground for the study of flows is the theory of ordinary differential equations. In particular, any (autonomous) differential equation $x' = F(x)$ with

L. Barreira, C. Valls, *Dynamical Systems*, Universitext, DOI 10.1007/978-1-4471-4835-7_1,

unique global solutions determines a flow. A natural continuation for the study of topics in the theory of flows, such as topological conjugacies, Lyapunov functions and stability theory, index theory, bifurcation theory, and Hamiltonian dynamics, would be our book [12] since the level and philosophy of the presentation are quite similar. We refer the reader to the books [3, 5, 21, 23] for additional topics in the theory of dynamical systems with continuous time.

1.1.2 Topological Dynamics

In Chap. 3 we consider the class of topological dynamical systems, that is, of continuous maps or homeomorphisms of a topological space. For example, one can ask whether there are dense orbits or whether all orbits are dense. One can also ask whether a given orbit returns arbitrarily close to itself, which corresponds the concept of recurrence. In particular, we study the notions of α-limit set and ω-limit set, which to some extent capture the asymptotic behavior of a dynamical system. We also consider several notions of topological recurrence, such as topological transitivity and topological mixing.

Then we introduce the notion of the topological entropy of a dynamical system (with discrete time), which is a measure of the complexity of a dynamical system from the topological point of view. We also illustrate its computation in various examples. Topological entropy and some of its modifications and generalizations stand as principal measures of the complexity of a dynamics, from various points of view. The topics in Sects. 3.4.3 and 3.4.4 on alternative characterizations of the topological entropy and on the particular case of expansive maps are somewhat more advanced.

For additional topics in topological dynamics, topological recurrence and topological entropy we refer the reader to the books [17, 18, 32, 53].

1.1.3 Low-Dimensional Dynamics

In Chap. 4 we consider several classes of dynamical systems in low-dimensional spaces. This essentially means dimension 1 for discrete time and dimension 2 for continuous time. In particular, we consider homeomorphisms and diffeomorphisms of the circle, continuous maps of the interval and flows defined by autonomous differential equations in the plane. For the orientation-preserving homeomorphisms of the circle, we introduce the notion of rotation number and we describe the behavior of the orbits depending on whether it is rational or irrational. We also study the existence of periodic points for the continuous maps of the interval and we establish Sharkovsky's theorem relating the existence of periodic points with different periods. Finally, we give a brief introduction to the Poincaré–Bendixson theory of flows on surfaces. The topics in Sects. 4.2 and 4.3.2 on diffeomorphisms of the circle with irrational rotation number that are topologically conjugate to rotations and on Sharkovsky's theorem are somewhat more advanced. For additional topics in low-dimensional dynamics we refer the reader to [2, 25, 32] for the case of discrete time and to [12, 23] for the case of continuous time.

In certain cases, the low-dimensionality of the space allows one to present some notions and results without the technical complications of arbitrary spaces. On the other hand, one should be aware that some of the methods and results only work precisely because of the low-dimensionality of the space. For example, the use of conformality for a smooth dynamics of the circle, which means that the derivative at each point is a multiple of an isometry, or the use of Jordan's curve theorem for a flow in the plane make some of the results in Chap. 4 belong strictly to low-dimensional dynamics.

1.1.4 Hyperbolic Dynamics

Chapter 5 is an introduction to hyperbolic dynamics, which can be described as the study of the properties of a smooth dynamics that expands or contracts in some privileged directions. All the pre-requisites from the theory of smooth manifolds are fully recalled in the chapter. After introducing the notion of a hyperbolic set, we describe the Smale horseshoe and some of its modifications. This allows us to illustrate some notions and results without the additional complications arising from considering arbitrary hyperbolic sets. We also establish the continuity of the stable and unstable spaces on the base point. Moreover, we discuss the characterization of hyperbolic sets in terms of invariant families of cones. In particular, this allows us to describe some stability properties of hyperbolic sets under sufficiently small perturbations. The relations between invariant families of cones, Lyapunov functions and hyperbolicity are in fact important in much more general classes of dynamics.

Chapter 6 is a natural continuation of Chap. 5 and considers topics that are somewhat less elementary. In particular, we describe the behavior of the orbits of a diffeomorphism near a hyperbolic fixed point, establishing two fundamental results of hyperbolic dynamics: the Grobman–Hartman theorem and the Hadamard–Perron theorem. The proofs are somewhat simple-minded but also long and unavoidably more technical. We also establish the existence of stable and unstable manifolds for all points of a hyperbolic set, with an elaboration of the proof of the Hadamard–Perron theorem, and we show how they give rise to a local product structure for any locally maximal hyperbolic set.

Section 6.3 is an introduction to the study of geodesic flows on surfaces of constant negative curvature and their hyperbolicity. All the pre-requisites from hyperbolic geometry are fully recalled in the section. In particular, we consider isometries, Möbius transformations, geodesics as the shortest paths between two points, quotients by isometries and the construction of compact surfaces of genus at least 2. The material showing that the geodesic flow on compact surfaces is hyperbolic is somewhat more advanced.

For additional topics in hyperbolic dynamics, such as transversality and genericity, homoclinic behavior, and growth of the number of periodic points, we refer the reader to the books [28, 31, 32, 44, 45, 55, 60]. See also [12, 21] for the case of continuous time.

We refer the reader to [4, 13, 34] for background and further developments of hyperbolic geometry.

1.1.5 Symbolic Dynamics

Chapter 7 is an introduction to symbolic dynamics, with emphasis on its relations to hyperbolic dynamics. In particular, we illustrate how one can associate a symbolic dynamics (also called a coding) to a hyperbolic set and how it can be used to solve certain problems related to the counting of periodic points (without actually finding them). The examples include expanding maps, quadratic maps and the Smale horseshoe. We also consider topological Markov chains, and their periodic points, topological entropy, recurrence properties and zeta functions.

For additional topics in symbolic dynamics and its relations to hyperbolic dynamics we refer the reader to the books [17, 32]. Good references for the core of symbolic dynamics are [36, 37]. See also [48] for a detailed study of zeta functions in hyperbolic dynamics.

1.1.6 Ergodic Theory

Chapter 8 is a first introduction to ergodic theory and the consequences of the existence of a finite invariant measure. After introducing the notions of a measurable map and of an invariant measure, we establish two basic but also fundamental results of ergodic theory: Poincaré's recurrence theorem and Birkhoff's ergodic theorem. We also introduce the notion of entropy and we illustrate its computation in various examples.

All the pre-requisites from measure theory and integration theory are fully recalled in the chapter. Nevertheless, due to the necessary familiarity with standard arguments of measure theory, the whole chapter should be considered more advanced. Further developments of ergodic theory clearly fall outside the scope of the book.

For additional topics in ergodic theory and its applications we refer the reader to the books [9, 38, 51, 61, 63, 65] and for further developments to [24, 47, 52, 56, 62]. In particular, [9, 65] include introductions to the thermodynamic formalism and its applications and [38, 52] include introductions to smooth ergodic theory (also called Pesin theory).

1.2 Further Topics and Suggested References

We emphasize that each of the topics discussed in the former section, that is, topological dynamics, low-dimensional dynamics, hyperbolic dynamics, symbolic dynamics and ergodic theory can be (and are) the object, by itself, of several books. So it would be impossible to be much more detailed in a self-contained introduction to the core of the theory of dynamical systems and choices had to be made. Certainly, our particular selection of topics may also reflect a personal taste. But it would be very hard to argue, perhaps with the exception of ergodic theory, that any of the topics considered in the book should be omitted from a first introduction.

The following is an incomplete list of topics that were left out of the book together with recommendations for further reading:

- holomorphic dynamics (see [14, 19, 39–41, 54]);
- bifurcation theory and normal forms (see [5, 12, 22, 28]);
- Hamiltonian dynamics (see [1, 6, 12, 32, 42]);
- dimension theory and multifractal analysis (see [7, 49]);
- thermodynamic formalism and its applications (see [8, 16, 35, 57]);
- hyperbolicity and homoclinic bifurcations (see [46]);
- partial hyperbolicity and stable ergodicity (see [50]);
- nonuniform hyperbolicity and smooth ergodic theory (see [10, 11, 15]);
- hyperbolic systems with singularities and billiards (see [20, 33]);
- algebraic dynamics and ergodic theory (see [58]);
- infinite-dimensional dynamics (see [29, 30, 59, 64]).

1.3 Suggestions for Courses Based on the Book

The book can be used as a basis for several advanced undergraduate or beginning graduate courses. Other than some basic pre-requisites from linear algebra, differential and integral calculus, complex analysis and topology, all the notions and results used in the book are recalled along the way.

The interdependence of the chapters is indicated in Fig. 1.1. An arrow going from Chapter A to Chapter B means that part of the material in Chapter A is used in Chapter B. This leads naturally to the following courses:

1. topological dynamics and symbolic dynamics: Chaps. 2, 3 and 7;
2. hyperbolic dynamics and symbolic dynamics: Chaps. 2, 5 and 7;
3. low-dimensional dynamics: Chaps. 2, 3 and 4;
4. hyperbolic dynamics, including geodesic flows: Chaps. 2, 5 and 6;
5. symbolic dynamics and ergodic theory: Chaps. 2, 7 and 8.

Other selections are also possible, depending on the audience and on the available time. Moreover, some sections can be used for short expositions, such as for example Sects. 3.4.3, 3.4.4, 4.2 and 4.3.2, and all sections in Chaps. 6 and 7.

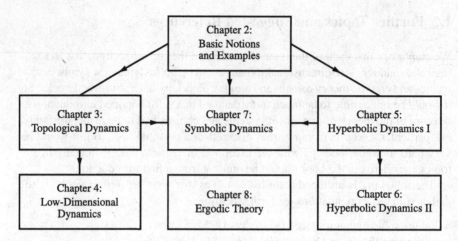

Fig. 1.1 Interdependence of the chapters

Chapter 2
Basic Notions and Examples

In this chapter we introduce the notion of a dynamical system, both for discrete and continuous time. We also describe many examples, including rotations and expanding maps of the circle, endomorphisms and automorphisms of the torus, and autonomous differential equations and their flows. Together with other examples introduced throughout the book, these are used to illustrate new concepts and results. We also describe some basic constructions determining new dynamical systems, including suspension flows and Poincaré maps. Finally, we consider the notion of an invariant set, both for maps and flows.

2.1 The Notion of a Dynamical System

In the case of discrete time, a dynamical system is simply a map.

Definition 2.1 Any map $f : X \to X$ is called a *dynamical system with discrete time* or simply a *dynamical system*.

We define recursively

$$f^{n+1} = f \circ f^n$$

for each $n \in \mathbb{N}$. We also write $f^0 = \mathrm{Id}$, where Id is the identity map. Clearly,

$$f^{m+n} = f^m \circ f^n \tag{2.1}$$

for every $m, n \in \mathbb{N}_0$, where $\mathbb{N}_0 = \mathbb{N} \cup \{0\}$. When the map f is invertible, we also define $f^{-n} = (f^{-1})^n$ for each $n \in \mathbb{N}$. In this case, identity (2.1) holds for every $m, n \in \mathbb{Z}$.

Example 2.1 Given dynamical systems $f : X \to X$ and $g : Y \to Y$, we define a new dynamical system $h : X \times Y \to X \times Y$ by

$$h(x, y) = \big(f(x), g(y)\big).$$

L. Barreira, C. Valls, *Dynamical Systems*, Universitext, DOI 10.1007/978-1-4471-4835-7_2,

We note that if f and g are invertible, then the map h is also invertible and its inverse is given by

$$h^{-1}(x, y) = \left(f^{-1}(x), g^{-1}(y)\right).$$

Now we consider the case of continuous time.

Definition 2.2 A family of maps $\varphi_t \colon X \to X$ for $t \geq 0$ such that $\varphi_0 = \mathrm{Id}$ and

$$\varphi_{t+s} = \varphi_t \circ \varphi_s \quad \text{for every } t, s \geq 0$$

is called a *semiflow*. A family of maps $\varphi_t \colon X \to X$ for $t \in \mathbb{R}$ such that $\varphi_0 = \mathrm{Id}$ and

$$\varphi_{t+s} = \varphi_t \circ \varphi_s \quad \text{for every } t, s \in \mathbb{R}$$

is called a *flow*.

We also say that a family of maps φ_t is a *dynamical system with continuous time* or simply a *dynamical system* if it is a flow or a semiflow. We note that if φ_t is a flow, then

$$\varphi_t \circ \varphi_{-t} = \varphi_{-t} \circ \varphi_t = \varphi_0 = \mathrm{Id}$$

and thus, each map φ_t is invertible and its inverse is given by $\varphi_t^{-1} = \varphi_{-t}$.

A simple example of a flow is any movement by translation with constant velocity.

Example 2.2 Given $y \in \mathbb{R}^n$, consider the maps $\varphi_t \colon \mathbb{R}^n \to \mathbb{R}^n$ defined by

$$\varphi_t(x) = x + ty, \quad t \in \mathbb{R}, \ x \in \mathbb{R}^n.$$

Clearly, $\varphi_0 = \mathrm{Id}$ and

$$\varphi_{t+s}(x) = x + (t+s)y$$
$$= (x + sy) + ty = (\varphi_t \circ \varphi_s)(x).$$

In other words, the family of maps φ_t is a flow.

Example 2.3 Given two flows $\varphi_t \colon X \to X$ and $\psi_t \colon Y \to Y$, for $t \in \mathbb{R}$, the family of maps $\alpha_t \colon X \times Y \to X \times Y$ defined for each $t \in \mathbb{R}$ by

$$\alpha_t(x, y) = \left(\varphi_t(x), \psi_t(y)\right)$$

is also a flow. Moreover,

$$\alpha_t^{-1}(x, y) = \left(\varphi_{-t}(x), \psi_{-t}(y)\right).$$

We emphasize that the expression *dynamical system* is used to refer both to dynamical systems with discrete time and to dynamical systems with continuous time.

2.2 Examples with Discrete Time

In this section we describe several examples of dynamical systems with discrete time.

2.2.1 Rotations of the Circle

We first consider the rotations of the circle. The *circle* S^1 is defined to be $\mathbb{R}/\mathbb{Z}$, that is, the real line with any two points $x, y \in \mathbb{R}$ identified if $x - y \in \mathbb{Z}$. In other words,

$$S^1 = \mathbb{R}/\mathbb{Z} = \mathbb{R}/\sim,$$

where $\sim$ is the equivalence relation on $\mathbb{R}$ defined by $x \sim y \Leftrightarrow x - y \in \mathbb{Z}$. The corresponding equivalence classes, which are the elements of S^1, can be written in the form

$$[x] = \{x + m : m \in \mathbb{Z}\}.$$

In particular, one can introduce the operations

$$[x] + [y] = [x + y] \quad \text{and} \quad [x] - [y] = [x - y].$$

One can also identify S^1 with $[0, 1]/\{0, 1\}$, where the endpoints of the interval $[0, 1]$ are identified.

Definition 2.3 Given $\alpha \in \mathbb{R}$, we define the *rotation* $R_\alpha : S^1 \to S^1$ by

$$R_\alpha([x]) = [x + \alpha]$$

(see Fig. 2.1).

Sometimes, we also write

$$R_\alpha(x) = x + \alpha \bmod 1,$$

thus identifying $[x]$ with its representative in the interval $[0, 1)$. The map R_α could also be called a *translation of the interval*. Clearly, $R_\alpha : S^1 \to S^1$ is invertible and its inverse is given by $R_\alpha^{-1} = R_{-\alpha}$.

Now we introduce the notion of a periodic point.

Definition 2.4 Given $q \in \mathbb{N}$, a point $x \in X$ is said to be a *q-periodic point* of a map $f : X \to X$ if $f^q(x) = x$. We also say that $x \in X$ is a *periodic point* of f if it is q-periodic for some $q \in \mathbb{N}$.

In particular, the fixed points, that is, the points $x \in X$ such that $f(x) = x$ are q-periodic for any $q \in \mathbb{N}$. Moreover, a q-periodic point is kq-periodic for any $k \in \mathbb{N}$.

Fig. 2.1 The rotation R_α

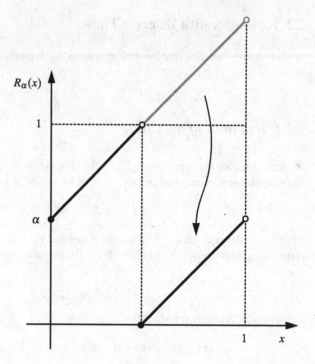

Definition 2.5 A periodic point is said to have *period q* if it is q-periodic but is not l-periodic for any $l < q$.

Now we consider the particular case of the rotations R_α of the circle. We verify that their behavior is very different depending on whether α is rational or irrational.

Proposition 2.1 *Given $\alpha \in \mathbb{R}$:*

1. *if $\alpha \in \mathbb{R} \setminus \mathbb{Q}$, then R_α has no periodic points;*
2. *if $\alpha = p/q \in \mathbb{Q}$ with p and q coprime, then all points of S^1 are periodic for R_α and have period q.*

Proof We note that $[x] \in S^1$ is q-periodic if and only if $[x + q\alpha] = [x]$, that is, if and only if $q\alpha \in \mathbb{Z}$. The two properties in the proposition follow easily from this observation.   □

2.2.2 Expanding Maps of the Circle

In this section we consider another family of maps of S^1.

Fig. 2.2 The expanding map E_2

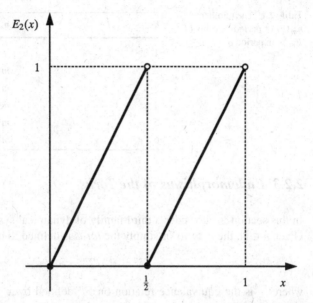

Definition 2.6 Given an integer $m > 1$, the *expanding map* $E_m \colon S^1 \to S^1$ is defined by

$$E_m(x) = mx \bmod 1.$$

For example, for $m = 2$, we have

$$E_2(x) = \begin{cases} 2x & \text{if } x \in [0, 1/2), \\ 2x - 1 & \text{if } x \in [1/2, 1) \end{cases}$$

(see Fig. 2.2).

Now we determine the periodic points of the expanding map E_m. Since $E_m^q(x) = m^q x \bmod 1$, a point $x \in S^1$ is q-periodic if and only if

$$m^q x - x = (m^q - 1)x \in \mathbb{Z}.$$

Hence, the q-periodic points of the expanding map E_m are

$$x = \frac{p}{m^q - 1}, \quad \text{for } p = 1, 2, \dots, m^q - 1. \tag{2.2}$$

Moreover, the number $n_m(q)$ of periodic points of E_m with period q can be computed easily for each given q (see Table 2.1 for $q \le 6$). For example, if q is prime, then

$$n_m(q) = m^q - m.$$

Table 2.1 The number $n_m(q)$ of periodic points of E_m with period q

q	$n_m(q)$
1	$m-1$
2	$m^2 - m = m^2 - 1 - (m-1)$
3	$m^3 - m = m^3 - 1 - (m-1)$
4	$m^4 - m^2 = m^4 - 1 - (m^2-1)$
5	$m^5 - m = m^5 - 1 - (m-1)$
6	$m^6 - m^3 - m^2 + m$

2.2.3 Endomorphisms of the Torus

In this section we consider a third family of dynamical systems with discrete time. Given $n \in \mathbb{N}$, the *n-torus* or simply the *torus* is defined to be

$$\mathbb{T}^n = \mathbb{R}^n / \mathbb{Z}^n = \mathbb{R}^n / \sim,$$

where $\sim$ is the equivalence relation on $\mathbb{R}^n$ defined by $x \sim y \Leftrightarrow x - y \in \mathbb{Z}^n$. The elements of $\mathbb{T}^n$ are thus the equivalence classes

$$[x] = \{x + y : y \in \mathbb{Z}^n\},$$

with $x \in \mathbb{R}^n$. Now let A be an $n \times n$ matrix with entries in $\mathbb{Z}$.

Definition 2.7 The *endomorphism of the torus* $T_A \colon \mathbb{T}^n \to \mathbb{T}^n$ is defined by

$$T_A([x]) = [Ax] \quad \text{for } [x] \in \mathbb{T}^n.$$

We also say that T_A is the *endomorphism of the torus induced by A*.

Since A is a linear transformation,

$$Ax - Ay \in \mathbb{Z}^n \quad \text{when } x - y \in \mathbb{Z}^n.$$

This shows that $Ay \in [Ax]$ when $y \in [x]$ and hence, T_A is well defined.

In general, the map T_A may not be invertible, even if the matrix A is invertible. When T_A is invertible, we also say that it is the *automorphism of the torus induced by A*. We represent in Fig. 2.3 the automorphism of the torus $\mathbb{T}^2$ induced by the matrix

$$A = \begin{pmatrix} 2 & 1 \\ 1 & 1 \end{pmatrix}.$$

Now we determine the periodic points of a class of automorphisms of the torus.

Proposition 2.2 Let $T_A \colon \mathbb{T}^n \to \mathbb{T}^n$ be an automorphism of the torus induced by a matrix A without eigenvalues with modulus 1. Then the periodic points of T_A are the points with rational coordinates, that is, the elements of $\mathbb{Q}^n / \mathbb{Z}^n$.

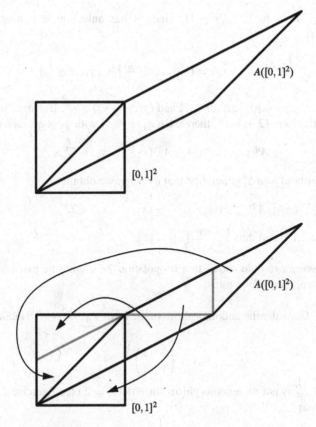

Fig. 2.3 An automorphism of the torus $\mathbb{T}^2$

Proof Let $[x] = [(x_1, \ldots, x_n)] \in \mathbb{T}^n$ be a periodic point. Then there exist $q \in \mathbb{N}$ and $y = (y_1, \ldots, y_n) \in \mathbb{Z}^n$ such that $A^q x = x + y$, that is,

$$\left(A^q - \mathrm{Id}\right)x = y.$$

Since A has no eigenvalues with modulus 1, the matrix $A^q - \mathrm{Id}$ is invertible and one can write

$$x = \left(A^q - \mathrm{Id}\right)^{-1} y.$$

Moreover, since $A^q - \mathrm{Id}$ has only integer entries, each entry of $(A^q - \mathrm{Id})^{-1}$ is a rational number and thus $x \in \mathbb{Q}^n$.

Now we assume that $[x] = [(x_1, \ldots, x_n)] \in \mathbb{Q}^n / \mathbb{Z}^n$ and we write

$$(x_1, \ldots, x_n) = \left(\frac{p_1}{r}, \ldots, \frac{p_n}{r}\right), \tag{2.3}$$

where $p_1, \ldots, p_n \in \{0, 1, \ldots, r - 1\}$. Since A has only integer entries, for each $q \in \mathbb{N}$ we have

$$A^q(x_1, \ldots, x_n) = \left(\frac{p'_1}{r}, \ldots, \frac{p'_n}{r}\right) + (y_1, \ldots, y_n)$$

for some $p'_1, \ldots, p'_n \in \{0, 1, \ldots, r - 1\}$ and $(y_1, \ldots, y_n) \in \mathbb{Z}^n$. But since the number of points of the form (2.3) is r^n, there exist $q_1, q_2 \in \mathbb{N}$ with $q_1 \neq q_2$ such that

$$A^{q_1}(x_1, \ldots, x_n) - A^{q_2}(x_1, \ldots, x_n) \in \mathbb{Z}^n.$$

Assuming, without loss of generality, that $q_1 > q_2$, we obtain

$$A^{q_1 - q_2}(x_1, \ldots, x_n) - (x_1, \ldots, x_n) \in \mathbb{Z}^n$$

(see Exercise 2.12) and thus $T_A^{q_1 - q_2}([x]) = [x]$.   □

The following example shows that Proposition 2.2 cannot be extended to arbitrary endomorphisms of the torus.

Example 2.4 Consider the endomorphism of the torus $T_A \colon \mathbb{T}^2 \to \mathbb{T}^2$ induced by the matrix

$$A = \begin{pmatrix} 3 & 1 \\ 1 & 1 \end{pmatrix}.$$

We note that T_A is not an automorphism since $\det A = 2$ (see Exercise 2.12). Now we observe that

$$T_A\left(0, \frac{1}{2}\right) = \left(\frac{1}{2}, \frac{1}{2}\right), \quad T_A\left(\frac{1}{2}, \frac{1}{2}\right) = (0, 0) \quad \text{and} \quad T_A(0, 0) = (0, 0).$$

This shows that the points with rational coordinates $(0, 1/2)$ and $(1/2, 1/2)$ are not periodic. On the other hand, the eigenvalues of A are $2 + \sqrt{2}$ and $2 - \sqrt{2}$, both without modulus 1.

2.3 Examples with Continuous Time

In this section we give some examples of dynamical systems with continuous time.

2.3.1 Autonomous Differential Equations

We first consider autonomous (ordinary) differential equations, that is, differential equations not depending explicitly on time. We verify that they give rise naturally to the concept of a flow.

Proposition 2.3 *Let* $f: \mathbb{R}^n \to \mathbb{R}^n$ *be a continuous function such that, given* $x_0 \in \mathbb{R}^n$*, the initial value problem*

$$\begin{cases} x' = f(x), \\ x(0) = x_0 \end{cases} \tag{2.4}$$

has a unique solution $x(t, x_0)$ *defined for* $t \in \mathbb{R}$*. Then the family of maps* $\varphi_t : \mathbb{R}^n \to \mathbb{R}^n$ *defined for each* $t \in \mathbb{R}$ *by*

$$\varphi_t(x_0) = x(t, x_0)$$

is a flow.

Proof Given $s \in \mathbb{R}$, consider the function $y: \mathbb{R} \to \mathbb{R}^n$ defined by

$$y(t) = x(t + s, x_0).$$

We have $y(0) = x(s, x_0)$ and

$$y'(t) = x'(t + s, x_0) = f(x(t + s, x_0)) = f(y(t))$$

for $t \in \mathbb{R}$. In other words, the function y is also a solution of the equation $x' = f(x)$. Since by hypothesis the initial value problem (2.4) has a unique solution, we obtain

$$y(t) = x(t, y(0)) = x(t, x(s, x_0)),$$

or equivalently,

$$x(t + s, x_0) = x(t, x(s, x_0)) \tag{2.5}$$

for $t, s \in \mathbb{R}$ and $x_0 \in \mathbb{R}^n$. It follows from (2.5) that $\varphi_{t+s} = \varphi_t \circ \varphi_s$. Moreover,

$$\varphi_0(x_0) = x(0, x_0) = x_0,$$

that is, $\varphi_0 = \mathrm{Id}$. This shows that the family of maps φ_t is a flow.   □

Now we consider two specific examples of autonomous differential equations and we describe the flows that they determine.

Example 2.5 Consider the differential equation

$$\begin{cases} x' = -y, \\ y' = x. \end{cases}$$

If $(x, y) = (x(t), y(t))$ is a solution, then

$$(x^2 + y^2)' = 2xx' + 2yy' = -2xy + 2yx = 0.$$

Thus, there exists a constant $r \geq 0$ such that

$$x(t)^2 + y(t)^2 = r^2.$$

Writing

$$x(t) = r\cos\theta(t) \quad \text{and} \quad y(t) = r\sin\theta(t),$$

where θ is some differentiable function, it follows from the identity $x' = -y$ that

$$-r\theta'(t)\sin\theta(t) = -r\sin\theta(t).$$

Hence, $\theta'(t) = 1$ and there exists a constant $c \in \mathbb{R}$ such that $\theta(t) = t + c$. Thus, writing

$$(x_0, y_0) = (r\cos c, r\sin c) \in \mathbb{R}^2,$$

we obtain

$$\begin{pmatrix} x(t) \\ y(t) \end{pmatrix} = \begin{pmatrix} r\cos(t+c) \\ r\sin(t+c) \end{pmatrix}$$

$$= \begin{pmatrix} \cos t \cdot r\cos c - \sin t \cdot r\sin c \\ \sin t \cdot r\cos c + \cos t \cdot r\sin c \end{pmatrix}$$

$$= \begin{pmatrix} \cos t & -\sin t \\ \sin t & \cos t \end{pmatrix} \begin{pmatrix} x_0 \\ y_0 \end{pmatrix}.$$

Notice that

$$R(t) = \begin{pmatrix} \cos t & -\sin t \\ \sin t & \cos t \end{pmatrix}$$

is a rotation matrix for each $t \in \mathbb{R}$. Since $R(0) = \text{Id}$, it follows from Proposition 2.3 that the family of maps $\varphi_t : \mathbb{R}^2 \to \mathbb{R}^2$ defined by

$$\varphi_t \begin{pmatrix} x_0 \\ y_0 \end{pmatrix} = R(t) \begin{pmatrix} x_0 \\ y_0 \end{pmatrix}$$

is a flow. Incidentally, the identity $\varphi_{t+s} = \varphi_t \circ \varphi_s$ is equivalent to the identity between rotation matrices

$$R(t+s) = R(t)R(s).$$

Example 2.6 Now we consider the differential equation

$$\begin{cases} x' = y, \\ y' = x. \end{cases}$$

If $(x, y) = (x(t), y(t))$ is a solution, then

$$(x^2 - y^2)' = 2xx' - 2yy' = 2xy - 2yx = 0.$$

Thus, there exists a constant $r \geq 0$ such that

$$x(t)^2 - y(t)^2 = r^2 \quad \text{or} \quad x(t)^2 - y(t)^2 = -r^2. \tag{2.6}$$

In the first case, one can write

$$x(t) = r\cosh\theta(t) \quad \text{and} \quad y(t) = r\sinh\theta(t),$$

where θ is some differentiable function. Since $x' = y$, we have

$$r\theta'(t)\sinh\theta(t) = r\sinh\theta(t)$$

and hence, $\theta(t) = t + c$ for some constant $c \in \mathbb{R}$. Thus, writing

$$(x_0, y_0) = (r\cosh c, r\sinh c) \in \mathbb{R}^2,$$

we obtain

$$
\begin{aligned}
\begin{pmatrix} x(t) \\ y(t) \end{pmatrix} &= \begin{pmatrix} r\cosh(t+c) \\ r\sinh(t+c) \end{pmatrix} \\
&= \begin{pmatrix} \cosh t \cdot r\cosh c + \sinh t \cdot r\sinh c \\ \sinh t \cdot r\cosh c + \cosh t \cdot \sinh c \end{pmatrix} \\
&= \begin{pmatrix} \cosh t & \sinh t \\ \sinh t & \cosh t \end{pmatrix} \begin{pmatrix} x_0 \\ y_0 \end{pmatrix} = S(t) \begin{pmatrix} x_0 \\ y_0 \end{pmatrix},
\end{aligned}
$$

where

$$S(t) = \begin{pmatrix} \cosh t & \sinh t \\ \sinh t & \cosh t \end{pmatrix}.$$

In the second case in (2.6), one can write

$$x(t) = r\sinh\theta(t) \quad \text{and} \quad y(t) = r\cosh\theta(t).$$

Proceeding analogously, we find that $\theta(t) = t + c$ for some constant $c \in \mathbb{R}$. Thus, writing

$$(x_0, y_0) = (r\sinh c, r\cosh c) \in \mathbb{R}^2,$$

we obtain

$$
\begin{aligned}
\begin{pmatrix} x(t) \\ y(t) \end{pmatrix} &= \begin{pmatrix} r\sinh(t+c) \\ r\cosh(t+c) \end{pmatrix} \\
&= \begin{pmatrix} \sinh t \cdot r\cosh c + \cosh t \cdot r\sinh c \\ \cosh t \cdot r\cosh c + \sinh t \cdot r\sinh c \end{pmatrix} \\
&= S(t) \begin{pmatrix} x_0 \\ y_0 \end{pmatrix}.
\end{aligned}
$$

Notice that $S(0) = \mathrm{Id}$. It follows from Proposition 2.3 that the family of maps $\psi_t \colon \mathbb{R}^2 \to \mathbb{R}^2$ defined by

$$\psi_t \begin{pmatrix} x_0 \\ y_0 \end{pmatrix} = S(t) \begin{pmatrix} x_0 \\ y_0 \end{pmatrix}$$

is a flow. In particular, it follows from the identity $\psi_{t+s} = \psi_t \circ \psi_s$ that

$$S(t+s) = S(t)S(s) \quad \text{for } t, s \in \mathbb{R}.$$

2.3.2 Discrete Time Versus Continuous Time

In this section we describe some relations between dynamical systems with discrete time and dynamical systems with continuous time.

Example 2.7 Given a flow $\varphi_t \colon X \to X$, for each $T \in \mathbb{R}$, the map $f = \varphi_T \colon X \to X$ is a dynamical system with discrete time. We note that f is invertible and that its inverse is given by $f^{-1} = \varphi_{-T}$. Similarly, given a semiflow $\varphi_t \colon X \to X$, for each $T \geq 0$, the map $f = \varphi_T \colon X \to X$ is a dynamical system with discrete time.

Now we describe a class of semiflows obtained from a dynamical system with discrete time $f \colon X \to X$. Given a function $\tau \colon X \to \mathbb{R}^+$, consider the set Y obtained from

$$Z = \bigl\{ (x, t) \in X \times \mathbb{R} : 0 \leq t \leq \tau(x) \bigr\}$$

identifying the points $(x, \tau(x))$ and $(f(x), 0)$, for each $x \in \mathbb{R}$. More precisely, we define $Y = Z/\!\sim$, where $\sim$ is the equivalence relation on Z defined by

$$(x, t) \sim (y, s) \quad \Leftrightarrow \quad y = f(x),\ t = \tau(x) \text{ and } s = 0$$

(see Fig. 2.4).

Definition 2.8 The *suspension semiflow* $\varphi_t \colon Y \to Y$ over f with *height* τ is defined for each $t \geq 0$ by

$$\varphi_t(x, s) = (x, s + t) \quad \text{when } s + t \in [0, \tau(x)] \tag{2.7}$$

(see Fig. 2.4).

One can easily verify that each suspension semiflow is indeed a semiflow. Moreover, when f is invertible, the family of maps φ_t in (2.7), for $t \in \mathbb{R}$, is a flow. It is called the *suspension flow* over f with height τ.

Conversely, given a semiflow $\varphi_t \colon Y \to Y$, sometimes one can construct a dynamical system with discrete time $f \colon X \to X$ such that the semiflow can be seen as a suspension semiflow over f.

Fig. 2.4 A suspension flow

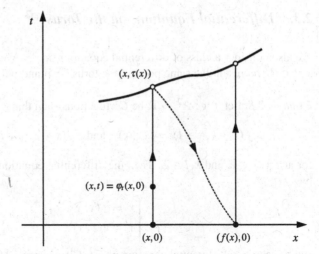

Fig. 2.5 A Poincaré section

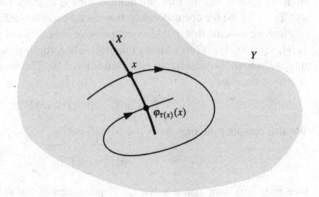

Definition 2.9 A set $X \subset Y$ is said to be a *Poincaré section* for a semiflow φ_t if

$$\tau(x) := \inf\{t > 0 : \varphi_t(x) \in X\} \in \mathbb{R}^+ \tag{2.8}$$

for each $x \in X$ (see Fig. 2.5), with the convention that $\inf \varnothing = +\infty$. The number $\tau(x)$ is called the *first return time* of x to the set X.

Thus, the first return time to X is a function $\tau : X \to \mathbb{R}^+$. We observe that (2.8) includes the hypothesis that each point of X returns to X. In fact, each point of X returns infinitely often to X.

Given a Poincaré section, one can introduce a corresponding Poincaré map.

Definition 2.10 Given a Poincaré section X for a semiflow φ_t, we define its *Poincaré map* $f : X \to X$ by

$$f(x) = \varphi_{\tau(x)}(x).$$

2.3.3 Differential Equations on the Torus $\mathbb{T}^2$

We also consider a class of differential equations on $\mathbb{T}^2$. We recall that two vectors $x, y \in \mathbb{R}^2$ represent the same point of the torus $\mathbb{T}^2$ if and only if $x - y \in \mathbb{Z}^2$.

Example 2.8 Let $f, g: \mathbb{R}^2 \to \mathbb{R}$ be C^1 functions such that

$$f(x+k, y+l) = f(x, y) \quad \text{and} \quad g(x+k, y+l) = g(x, y)$$

for any $x, y \in \mathbb{R}$ and $k, l \in \mathbb{Z}$. Then the differential equation in the plane $\mathbb{R}^2$ given by

$$\begin{cases} x' = f(x, y), \\ y' = g(x, y) \end{cases} \tag{2.9}$$

can be seen as a differential equation on $\mathbb{T}^2$. Clearly, Eq. (2.9) has unique solutions (that are global, that is, they are defined for $t \in \mathbb{R}$ since the torus is compact). Let $\varphi_t : \mathbb{T}^2 \to \mathbb{T}^2$ be the corresponding flow (see Proposition 2.3).

Now we assume that f takes only positive values. Then each solution $\varphi_t(0, z) = (x(t), y(t))$ of Eq. (2.9) crosses infinitely often the line segment $x = 0$, which is thus a *Poincaré section* for φ_t (see Definition 2.9). The first intersection (for $t > 0$) occurs at the time

$$T_z = \inf\{t > 0 : x(t) = 1\}.$$

We also consider the map $h: S^1 \to S^1$ defined by

$$h(z) = y(T_z) \tag{2.10}$$

(see Fig. 2.6). One can use the C^1 dependence of the solutions of a differential equation on the initial conditions to show that h is a diffeomorphism, that is, a bijective (one-to-one and onto) differentiable map with differentiable inverse (see Exercise 2.20).

For example, if $f = 1$ and $g = \alpha \in \mathbb{R}$, then

$$\varphi_t(0, z) = (t, z + t\alpha) \bmod 1.$$

Thus, $T_z = 1$ for each $z \in \mathbb{R}$ and

$$h(z) = z + \alpha \bmod 1 = R_\alpha(z).$$

2.4 Invariant Sets

In this section we introduce the notion of an invariant set with respect to a dynamical system.

Fig. 2.6 The Poincaré map determined by the Poincaré section $x = 0$

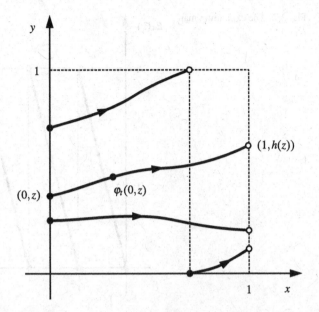

Definition 2.11 Given a map $f \colon X \to X$, a set $A \subset X$ is said to be:

1. *f-invariant* if $f^{-1}A = A$, where

$$f^{-1}A = \{x \in X : f(x) \in A\};$$

2. *forward f-invariant* if $f(A) \subset A$;
3. *backward f-invariant* if $f^{-1}A \subset A$.

Example 2.9 Consider the rotation $R_\alpha \colon S^1 \to S^1$. For $\alpha \in \mathbb{Q}$, each set

$$\gamma(x) = \{R_\alpha^n(x) : n \in \mathbb{Z}\}$$

is finite and R_α-invariant. More generally, if $\alpha \in \mathbb{Q}$, then a nonempty set $A \subset X$ is R_α-invariant if and only if it is a union of sets of the form $\gamma(x)$ (see the discussion after Definition 2.12). For example, the set $\mathbb{Q}/\mathbb{Z}$ is R_α-invariant.

On the other hand, for $\alpha \in \mathbb{R} \setminus \mathbb{Q}$, each set $\gamma(x)$ is also R_α-invariant, but now it is infinite. Again, a nonempty set $A \subset X$ is R_α-invariant if and only if it is a union of sets of the form $\gamma(x)$. One can show that each set $\gamma(x)$ is dense in S^1 (see Example 3.2) and thus, the closed R_α-invariant sets are $\varnothing$ and S^1.

Example 2.10 Now we consider the expanding map $E_4 \colon S^1 \to S^1$, given by

$$E_4(x) = \begin{cases} 4x & \text{if } x \in [0, 1/4), \\ 4x - 1 & \text{if } x \in [1/4, 2/4), \\ 4x - 2 & \text{if } x \in [2/4, 3/4), \\ 4x - 3 & \text{if } x \in [3/4, 1) \end{cases}$$

Fig. 2.7 The expanding map E_4

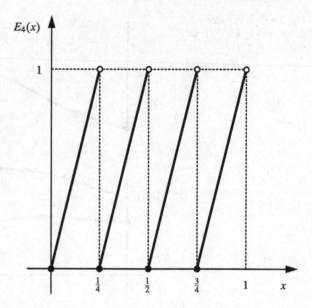

(see Fig. 2.7). For example, the set

$$A = \bigcap_{n \geq 0} E_4^{-n}\left([0, 1/4] \cup [2/4, 3/4]\right) \tag{2.11}$$

is forward E_4-invariant. We note that A is a *Cantor set*, that is, A is a closed set without isolated points and containing no intervals.

We also introduce the notions of orbit and semiorbit.

Definition 2.12 For a map $f : X \to X$, given a point $x \in X$, the set

$$\gamma^+(x) = \gamma_f^+(x) = \left\{ f^n(x) : n \in \mathbb{N}_0 \right\}$$

is called the *positive semiorbit* of x. Moreover, when f is invertible,

$$\gamma^-(x) = \gamma_f^-(x) = \left\{ f^{-n}(x) : n \in \mathbb{N}_0 \right\}$$

is called the *negative semiorbit* of x and

$$\gamma(x) = \gamma_f(x) = \left\{ f^n(x) : n \in \mathbb{Z} \right\}$$

is called the *orbit* of x.

We note that when f is invertible, a nonempty set $A \subset X$ is f-invariant if and only if it is a union of orbits. Indeed, $A \subset X$ is f-invariant if and only if

$$x \in A \quad \Leftrightarrow \quad x \in f^{-1}A \quad \Leftrightarrow \quad f(x) \in A.$$

By induction, this is equivalent to

$$x \in A \quad \Leftrightarrow \quad \{f^n(x) : n \in \mathbb{Z}\} \subset A \quad \Leftrightarrow \quad \gamma(x) \in A$$

since f is invertible. Thus, a nonempty set $A \subset X$ is f-invariant if and only if

$$A = \bigcup_{x \in A} \gamma(x).$$

Now we introduce the notion of an invariant set with respect to a flow or a semi-flow.

Definition 2.13 Given a flow $\Phi = (\varphi_t)_{t \in \mathbb{R}}$ of X, a set $A \subset X$ is said to be Φ-*invariant* if

$$\varphi_t^{-1} A = A \quad \text{for } t \in \mathbb{R}.$$

Given a semiflow $\Phi = (\varphi_t)_{t \geq 0}$ of X, a set $A \subset X$ is said to be Φ-*invariant* if

$$\varphi_t^{-1} A = A \quad \text{for } t \geq 0.$$

In the case of flows, since $\varphi_t^{-1} = \varphi_{-t}$ for $t \in \mathbb{R}$, a set $A \subset X$ is Φ-invariant if and only if

$$\varphi_t(A) = A \quad \text{for } t \in \mathbb{R}.$$

Example 2.11 Consider the differential equation

$$\begin{cases} x' = 2y^3, \\ y' = -3x. \end{cases} \tag{2.12}$$

Each solution $(x, y) = (x(t), y(t))$ satisfies

$$(3x^2 + y^4)' = 6xx' + 4y^3 y'$$
$$= 12xy^3 - 12y^3 x = 0.$$

Thus, for each set $I \subset \mathbb{R}^+$, the union

$$A = \bigcup_{a \in I} \{(x, y) \in \mathbb{R}^2 : 3x^2 + y^4 = a\}$$

is invariant with respect to the flow determined by Eq. (2.12).

We also introduce the notions of orbit and semiorbit for a semiflow.

Definition 2.14 For a semiflow $\Phi = (\varphi_t)_{t \geq 0}$ of X, given a point $x \in X$, the set

$$\gamma^+(x) = \gamma_\Phi^+(x) = \{\varphi_t(x) : t \geq 0\}$$

is called the *positive semiorbit* of x. Moreover, for a flow $\Phi = (\varphi_t)_{t \in \mathbb{R}}$ of X,

$$\gamma^-(x) = \gamma_\Phi^-(x) = \{\varphi_{-t}(x) : t \geq 0\}$$

is called the *negative semiorbit* of x and

$$\gamma(x) = \gamma_\Phi(x) = \{\varphi_t(x) : t \in \mathbb{R}\}$$

is called the *orbit* of x.

2.5 Exercises

Exercise 2.1 Determine whether the map $f \colon \mathbb{R} \to \mathbb{R}$ given by $f(x) = 3x - 3x^2$ has periodic points with period 2.

Exercise 2.2 Determine whether the map $f \colon \mathbb{R} \to \mathbb{R}$ given by $f(x) = x^2 + 1$ has periodic points with period 5.

Exercise 2.3 Given a continuous function $f \colon \mathbb{R} \to \mathbb{R}$, show that:

1. if $[a, b] \subset f([a, b])$, then f has a fixed point in $[a, b]$;
2. if $[a, b] \supset f([a, b])$, then f has a fixed point in $[a, b]$.

Exercise 2.4 Let $f \colon \mathbb{R} \to \mathbb{R}$ be a continuous function and let $[a, b]$ and $[c, d]$ be intervals in $\mathbb{R}$ such that

$$[c, d] \subset f([a, b]), \quad [a, b] \subset f([c, d]) \quad \text{and} \quad [a, b] \cap [c, d] = \varnothing.$$

Show that f has a periodic point with period 2.

Exercise 2.5 Show that if $f \colon [a, b] \to [a, b]$ is a homeomorphism (that is, a continuous bijective function with continuous inverse), then f has no periodic points with period 3 or larger.

Exercise 2.6 Determine whether there exists a homeomorphism $f \colon \mathbb{R} \to \mathbb{R}$ with:

1. a periodic point with period 2;
2. a periodic point with period 3.

Exercise 2.7 Show that any power of an expanding map is still an expanding map.

Exercise 2.8 Show that the set of periodic points of the expanding map E_m is dense in S^1.

Exercise 2.9 For each $q \in \mathbb{N}$, find the number of q-periodic points of the map $f \colon R \to R$ defined by $f(z) = z^2$ in the set

$$R = \left\{ z \in \mathbb{C} : |z| = 1 \right\}.$$

Exercise 2.10 Show that the number of periodic points of the expanding map E_m with period $p = q^r$, for q prime and $r \in \mathbb{N}$, is given by

$$n_m(p) = m^p - m^{p/q}.$$

Exercise 2.11 Find the smallest E_3-invariant set containing $[0, 1/3] \cup [2/3, 1]$.

Exercise 2.12 Show that the following properties are equivalent:

1. the endomorphism of the torus $T_A \colon \mathbb{T}^n \to \mathbb{T}^n$ is invertible;
2. $x \in \mathbb{Z}^n$ if and only if $Ax \in \mathbb{Z}^n$;
3. $|\det A| = 1$.

Exercise 2.13 Let $T_A \colon \mathbb{T}^n \to \mathbb{T}^n$ be an endomorphism of the torus. Show that for each $x \in \mathbb{Q}^n/\mathbb{Z}^n$, there exists an $m \in \mathbb{N}$ such that $T_A^m(x)$ is a periodic point of T_A.

Exercise 2.14 Show that the complement of a forward f-invariant set is backward f-invariant.

Exercise 2.15 Given a map $f \colon X \to X$, show that:

1. a set $A \subset X$ is f-invariant if and only if $f^{-1}A \subset A$ and $f(A) \subset A$;
2. a set $A \subset X$ is f-invariant if and only if $X \setminus A$ is f-invariant.

Exercise 2.16 Show that if X is a Poincaré section for a semiflow φ_t, then:

1. φ_t has no fixed points in X;
2. f is invertible when φ_t is a flow.

Exercise 2.17 Find the flow determined by the equation $x'' + 4x = 0$.

Exercise 2.18 Find the flow determined by the equation $x'' - 5x' + 6x = 0$.

Exercise 2.19 Show that the equation $x' = x^2$ does not determine a flow.

Exercise 2.20 Use the C^1 dependence of the solutions of a differential equation on the initial conditions[1] together with the implicit function theorem to show that the map h defined by (2.10) is a diffeomorphism.

---

[1] **Theorem** (See for example [12]) If $f \colon D \to \mathbb{R}^n$ is a C^1 function in an open set $D \subset \mathbb{R}^n$ and $\varphi(\cdot, x_0)$ is the solution of the initial value problem (2.4), then the function $(t, x) \mapsto \varphi(t, x)$ is of class C^1.

Chapter 3
Topological Dynamics

In this chapter we consider the class of topological dynamical systems, that is, the class of continuous maps of a topological space X. For simplicity of the exposition, we always assume that X is a locally compact metric space with a countable basis (this means, respectively, that each point has a compact neighborhood and that there exists a countable family of open sets such that each open set can be written as a union of elements of this family). In particular, we consider the notions of α-limit set and ω-limit set, as well as some basic notions and results of (topological) recurrence, including the notions of topological transitivity and topological mixing. Finally, we introduce the notion of topological entropy, which measures the complexity of a dynamical system, and we illustrate its computation in several examples. We also show that the topological entropy is a topological invariant and we give several alternative characterizations, including for expansive maps.

3.1 Topological Dynamical Systems

In this section we introduce the notion of a topological dynamical system.

Definition 3.1 A continuous map $f : X \to X$ is said to be a *topological dynamical system with discrete time* or simply a *topological dynamical system*. When f is a homeomorphism (that is, a bijective continuous map with continuous inverse), we also say that it is an *invertible topological dynamical system*.

For example, each rotation $R_\alpha : S^1 \to S^1$ is a homeomorphism of the circle, with the topology and the distance on $S^1 = \mathbb{R}/\mathbb{Z}$ induced from $\mathbb{R}$. More precisely, the topology of S^1 is generated by the sets of the form (a, b) and $[0, a) \cup (b, 1]$, with $0 < a < b < 1$, and the distance d on S^1 is given by

$$d(x, y) = \min\{|(x + k) - (y + l)| : k, l \in \mathbb{Z}\}$$
$$= \min\{|x - y - m| : m \in \mathbb{Z}\}. \tag{3.1}$$

L. Barreira, C. Valls, *Dynamical Systems*, Universitext, DOI 10.1007/978-1-4471-4835-7_3,   27
© Springer-Verlag London 2013

Now we consider the case of continuous time.

Definition 3.2 Any flow (respectively, any semiflow) $\varphi_t \colon X \to X$ such that the map $(t, x) \mapsto \varphi_t(x)$ is continuous in $\mathbb{R} \times X$ (respectively, in $\mathbb{R}_0^+ \times X$) is said to be a *topological flow* (respectively, a *topological semiflow*). Any topological flow or semiflow is also said to be a *topological dynamical system with continuous time* or simply a *topological dynamical system*.

In particular, the continuity assumptions imply that each map $\varphi_t \colon X \to X$ is continuous (in the case of flows it is even a homeomorphism).

Example 3.1 Let $f \colon \mathbb{R}^n \to \mathbb{R}^n$ be a Lipschitz function with $f(0) = 0$. We recall that f is said to be a *Lipschitz function* if there exists an $L > 0$ such that

$$\|f(x) - f(y)\| \le L\|x - y\|$$

for $x, y \in \mathbb{R}^n$. Now we consider the initial value problem (2.4), which has a unique solution $x(t, x_0)$ for each $x_0 \in \mathbb{R}^n$. It follows from

$$x(t, x_0) = x_0 + \int_0^t f(x(s, x_0))\, ds$$

that

$$\|x(t, x_0)\| \le \|x_0\| + \left| \int_0^t \|f(x(s, x_0))\|\, ds \right|$$

$$\le \|x_0\| + L \left| \int_0^t \|x(s, x_0)\|\, ds \right|.$$

By Gronwall's lemma,[1] we obtain

$$\|x(t, x_0)\| \le \|x_0\| e^{L|t|}$$

for t in the domain of the solution. This implies that the solution $\varphi_t(x_0) = x(t, x_0)$ is defined for $t \in \mathbb{R}$. It follows from the continuous dependence of the solutions

---

[1]**Theorem** (See for example [12]) If $u, v \colon [a, b] \to \mathbb{R}$ are continuous functions with $v \ge 0$ such that

$$u(t) \le c + \int_a^t u(s)v(s)\, ds \quad \text{for } t \in [a, b],$$

then

$$u(t) \le c \exp \int_a^t v(s)\, ds \quad \text{for } t \in [a, b].$$

of a differential equation on the initial conditions[2] that the flow $\varphi_t \colon \mathbb{R}^n \to \mathbb{R}^n$ is a topological dynamical system.

3.2 Limit Sets and Basic Properties

In this section we introduce the notions of α-limit set and ω-limit set for a dynamical system. These sets contain information about the asymptotic behavior of each orbit. More precisely, the ω-limit set of a point x is formed by the points that are arbitrarily approximated by the images $f^n(x)$ while the α-limit set of x is formed by the points that are arbitrarily approximated by the preimages $f^{-n}(x)$.

3.2.1 Discrete Time

We begin with the case of discrete time. Let $f \colon X \to X$ be a map (it need not be continuous).

Definition 3.3 Given a point $x \in X$, the *ω-limit set* of x is defined by

$$\omega(x) = \omega_f(x) = \bigcap_{n \in \mathbb{N}} \overline{\{f^m(x) : m \geq n\}}.$$

Moreover, when f is invertible, the *α-limit set* of x is defined by

$$\alpha(x) = \alpha_f(x) = \bigcap_{n \in \mathbb{N}} \overline{\{f^{-m}(x) : m \geq n\}}.$$

Now we give some examples.

Example 3.2 Let $R_\alpha \colon S^1 \to S^1$ be a rotation of the circle. For $\alpha \in \mathbb{Q}$, we have

$$\omega(x) = \alpha(x) = \gamma(x)$$

for $x \in S^1$. On the other hand, for $\alpha \in \mathbb{R} \setminus \mathbb{Q}$, we have

$$\omega(x) = \alpha(x) = S^1 \tag{3.2}$$

for $x \in S^1$. In order to establish (3.2), we show that the sets

$$\{R_\alpha^m(x) : m \geq n\} \quad \text{and} \quad \{R_\alpha^{-m}(x) : m \geq n\} \tag{3.3}$$

---

[2]**Theorem** (See for example [12]) If $f \colon D \to \mathbb{R}^n$ is a Lipschitz function on an open set $D \subset \mathbb{R}^n$ and $\varphi(\cdot, x_0)$ is the solution of the initial value problem (2.4), then the function $(t, x) \mapsto \varphi(t, x)$ is continuous.

are dense in S^1 for every $x \in S^1$ and $n \in \mathbb{N}$. We first assume that $R_\alpha^{m_1}(x) = R_\alpha^{m_2}(x)$ for some integers $m_1 > m_2 \geq n$. This is the same as

$$x + m_1\alpha = x + m_2\alpha \mod 1$$

or equivalently,

$$m_1\alpha - m_2\alpha = m$$

for some $m \in \mathbb{Z}$. Then $\alpha = m/(m_1 - m_2)$, but this is impossible since α is irrational. Thus, for each $n \in \mathbb{N}$, the points $R_\alpha^m(x)$ are pairwise distinct for $m \geq n$. Now let us take $\varepsilon > 0$ and $N \in \mathbb{N}$ such that $1/N < \varepsilon$. Since the points $R_\alpha^n(x), R_\alpha^{n+1}(x), \ldots, R_\alpha^{n+N}(x)$ are distinct, there exist integers i_1 and i_2 such that $0 \leq i_1 < i_2 \leq N$ and

$$d\left(R_\alpha^{n+i_1}(x), R_\alpha^{n+i_2}(x)\right) \leq \frac{1}{N} < \varepsilon, \tag{3.4}$$

where d is the distance in (3.1). Hence,

$$d\left(R_\alpha^{i_2-i_1}(x), x\right) = d\left(R_\alpha^{i_2-i_1}\left(R_\alpha^{n+i_1}(x)\right), R_\alpha^{n+i_1}(x)\right)$$
$$= d\left(R_\alpha^{n+i_2}(x), R_\alpha^{n+i_1}(x)\right) < \varepsilon$$

and the sequence $x_m = R_\alpha^{m(i_2-i_1)}(x)$, with $m \in \mathbb{N}$, is ε-dense in S^1 (in other words, for each $y \in S^1$ there exists an $m \in \mathbb{N}$ such that $d(y, x_m) < \varepsilon$). Since ε is arbitrary, we conclude that the first set in (3.3) is dense in S^1. In order to prove that the second set is also dense in S^1 it is sufficient to repeat the above argument to show that there exist no integers $m_1 > m_2 \geq n$ with $R_\alpha^{-m_1}(x) = R_\alpha^{-m_2}(x)$ (or to observe that this identity is equivalent to $R_\alpha^{m_1}(x) = R_\alpha^{m_2}(x)$).

Example 3.3 Given $\alpha \in \mathbb{R} \setminus \mathbb{Q}$ and $\delta > 0$, we show that there exist integers $p \in \mathbb{Z}$ and $q \in (0, 1/\delta]$ such that

$$\left|\alpha - \frac{p}{q}\right| \leq \frac{\delta}{q}. \tag{3.5}$$

Take an integer $N > 1$ such that $1/N \leq \delta$. Proceeding as in (3.4), we find that there exist integers m and n such that $0 \leq n < m \leq N$ and

$$d\left(R_\alpha^m(0), R_\alpha^n(0)\right) < \frac{1}{N}.$$

Taking $q = m - n$, we obtain

$$d\left(R_\alpha^q(0), 0\right) = d\left(R_\alpha^q\left(R_\alpha^n(0)\right), R_\alpha^n(0)\right)$$
$$= d\left(R_\alpha^m(0), R_\alpha^n(0)\right) < \frac{1}{N} \leq \delta.$$

Finally, it follows from (3.1) that there exists a $p \in \mathbb{Z}$ such that

$$\left| R_\alpha^q(0) - p \right| < \frac{1}{N} \le \delta.$$

Since $1/N < 1$ and $R_\alpha^q(0) = q\alpha \bmod 1$, we obtain

$$|q\alpha - p| < \frac{1}{N} \le \delta,$$

which establishes inequality (3.5).

Example 3.4 Now we consider the expanding map $E_2 \colon S^1 \to S^1$ and the point

$$x = 0.\boxed{0}\,\boxed{1}\,\boxed{00}\,\boxed{01}\,\boxed{10}\,\boxed{11}\,\boxed{000}\,\boxed{001}\,\boxed{010}\cdots,$$

whose base-2 expansion comprises the sequence of all length 1 binary strings $(0, 1)$ followed by all length 2 binary strings $(00, 01, 10, 11)$, then all length 3 binary strings $(000, 001, 010, \ldots)$, and so on. Since

$$E_2^m(0.x_1 x_2 \cdots) = 0.x_{m+1} x_{m+2} \cdots,$$

each set $\{E_2^m(x) : m \ge n\}$ is dense in S^1 and thus $\omega(x) = S^1$.

We note that the same happens when x is replaced by any point in S^1 whose base-2 representation contains all finite binary strings, in any order.

Example 3.5 Let $f \colon \mathbb{R}^2 \to \mathbb{R}^2$ be the map given by

$$f(r\cos\theta, r\sin\theta) = \left(\frac{r}{r + (1-r)/2} \cos\left(\theta + \frac{\pi}{4}\right), \frac{r}{r + (1-r)/2} \sin\left(\theta + \frac{\pi}{4}\right) \right)$$

(see Fig. 3.1). One can easily verify that f is invertible and that

$$f^n(r\cos\theta, r\sin\theta)$$
$$= \left(\frac{r}{r + (1-r)/2^n} \cos\left(\theta + \frac{n\pi}{4}\right), \frac{r}{r + (1-r)/2^n} \sin\left(\theta + \frac{n\pi}{4}\right) \right)$$

for each $n \in \mathbb{Z}$. Clearly, the origin ($r = 0$) and the circle $r = 1$ are f-invariant sets. For $r > 0$, we have

$$\lim_{n \to +\infty} \frac{r}{r + (1-r)/2^n} = 1$$

and thus, the ω-limit set of a point $p = (r\cos\theta, r\sin\theta)$ outside the origin is given by

$$\omega(p) = \left\{ \left(\cos\left(\theta + \frac{n\pi}{4}\right), \sin\left(\theta + \frac{n\pi}{4}\right) \right) : n = 0, 1, 2, 3, 4, 5, 6, 7 \right\}.$$

Fig. 3.1 The map f in
Example 3.5

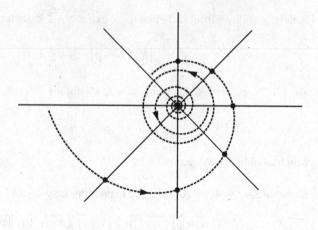

On the other hand, for $r \in (0, 1)$, we have

$$\lim_{n \to -\infty} \frac{r}{r + (1 - r)/2^n} = 0$$

and thus, the α-limit set of any point in the region $0 < r < 1$ is the origin.

Now we establish some properties of the α-limit sets and ω-limit sets. We recall that X is a metric space, say with distance d.

Proposition 3.1 *Given a map $f : X \to X$, for each $x \in X$ the following properties hold:*

1. *$y \in \omega(x)$ if and only if there exists a sequence $n_k \nearrow \infty$ in $\mathbb{N}$ such that $f^{n_k}(x) \to y$ when $k \to \infty$;*
2. *if f is continuous, then $\omega(x)$ is forward f-invariant.*

Proof We have $\omega(x) = \bigcap_{m \geq 1} \overline{A_m}$, where

$$A_m = \{ f^n(x) : n \geq m \}.$$

Now let $y \in \omega(x)$. We consider two cases:

1. if $y \notin \bigcap_{m \geq 1} A_m$, then there exists $p \geq 1$ such that $y \notin A_p$. Hence, $y \in \overline{A_p} \setminus A_p$ and there exists a sequence $n_k \nearrow \infty$ in $\mathbb{N}$ such that $f^{n_k}(x) \to y$ when $k \to \infty$.
2. if $y \in \bigcap_{m \geq 1} A_m$, then there exists $p \geq 1$ such that $y = f^p(x)$. Since $y \in A_m$ for $m > p$, there exists $q > p$ such that $y = f^q(x)$. Thus,

$$f^{(q-p)k}\big(f^p(x)\big) = y \quad \text{for } k \in \mathbb{N}$$

and the increasing sequence $n_k = (q - p)k + p$ satisfies $f^{n_k}(x) = y$.

On the other hand, if there exists a sequence $n_k \nearrow \infty$ in $\mathbb{N}$ such that $f^{n_k}(x) \to y$ when $k \to \infty$, then $y \in \overline{A_m}$ for every $m \in \mathbb{N}$. Hence, $y \in \omega(x)$.

Now let us take $y \in \omega(x)$ and $n \in \mathbb{N}$. By the first property, there exists a sequence $n_k \nearrow \infty$ in $\mathbb{N}$ such that $f^{n_k}(x) \to y$ when $k \to \infty$. It follows from the continuity of f that $f^{n_k+n}(x) \to f^n(y)$ when $k \to \infty$ and hence $f^n(y) \in \omega(x)$. This shows that $\omega(x)$ is forward f-invariant.   □

Proposition 3.2 *Given a continuous map $f: X \to X$, if the positive semiorbit $\gamma^+(x)$ of a point $x \in X$ has compact closure, then:*

1. *$\omega(x)$ is compact and nonempty;*
2. *$\inf\{d(f^n(x), y) : y \in \omega(x)\} \to 0$ when $n \to \infty$.*

Proof For the first property, we note that by definition the set $\omega(x)$ is closed. Since $\omega(x) \subset \overline{\gamma^+(x)}$ and the closure of the semiorbit $\gamma^+(x)$ is compact, the set $\omega(x)$ is also compact.

Now we consider the sequence $f^n(x)$. Since it is contained in the compact subset $\overline{\gamma^+(x)}$ of the metric space X, there exists a convergent subsequence $f^{n_k}(x)$, with $n_k \nearrow \infty$ when $k \to \infty$. Thus, one can apply the first property in Proposition 3.1 to conclude that the limit of $f^{n_k}(x)$ is in $\omega(x)$. This shows that $\omega(x)$ is nonempty.

Finally, if the last property did not hold, then there would exist $\delta > 0$ and a sequence $n_k \nearrow \infty$ such that

$$\inf\{d(f^{n_k}(x), y) : y \in \omega(x)\} \geq \delta \qquad (3.6)$$

for $k \in \mathbb{N}$. Since the set $\overline{\gamma^+(x)}$ is compact, there would exist a convergent subsequence $f^{m_k}(x)$ of $f^{n_k}(x)$ whose limit, by the first property in Proposition 3.1, is a point $p \in \omega(x)$. On the other hand, it follows from (3.6) that

$$d(f^{m_k}(x), y) \geq \delta$$

for $k \in \mathbb{N}$ and $y \in \omega(x)$ and thus, $d(p, y) \geq \delta$ for $y \in \omega(x)$. But this is impossible since $p \in \omega(x)$. This contradiction yields the last property in the proposition.   □

For invertible maps, we have the following results for the α-limit set.

Proposition 3.3 *Given an invertible map $f: X \to X$, for each $x \in X$ the following properties hold:*

1. *$y \in \alpha(x)$ if and only if there exists a sequence $n_k \nearrow \infty$ in $\mathbb{N}$ such that $f^{-n_k}(x) \to y$ when $k \to \infty$;*
2. *if f has a continuous inverse, then $\alpha(x)$ is backward f-invariant.*

Proposition 3.4 *Given an invertible map $f: X \to X$ with continuous inverse, if the negative semiorbit $\gamma^-(x)$ of a point $x \in X$ has compact closure, then:*

1. $\alpha(x)$ *is compact and nonempty;*
2. $\inf\{d(f^n(x), y) : y \in \alpha(x)\} \to 0$ *when* $n \to -\infty$.

In order to obtain these two propositions, it suffices to apply Propositions 3.1 and 3.2 to the map $g = f^{-1}$.

3.2.2 Continuous Time

Now we introduce the notions of α-limit set and ω-limit set for a dynamical system with continuous time.

Definition 3.4 Given a semiflow $\Phi = (\varphi_t)_{t \geq 0}$ of X, the *ω-limit set* of a point $x \in X$ is defined by

$$\omega(x) = \omega_\Phi(x) = \bigcap_{t>0} \overline{\{\varphi_s(x) : s > t\}}.$$

Moreover, given a flow $\Phi = (\varphi_t)_{t \in \mathbb{R}}$ of X, the *α-limit set* of a point $x \in X$ is defined by

$$\alpha(x) = \alpha_\Phi(x) = \bigcap_{t<0} \overline{\{\varphi_s(x) : s < t\}}.$$

Example 3.6 Consider the differential equation in polar coordinates

$$\begin{cases} r' = r(r-1)(r-2), \\ \theta' = 1. \end{cases} \tag{3.7}$$

We note that $r' > 0$ for $r \in (0, 1) \cup (2, +\infty)$ and that $r' < 0$ for $r \in (1, 2)$. Now we consider the sets

$$C_r = \{(x, y) \in \mathbb{R}^2 : x^2 + y^2 = r^2\}$$

for $r > 0$. Given $p \in C_r$, we have

$$\alpha(p) = \omega(p) = \{(0, 0)\} \quad \text{for } r = 0,$$

$$\alpha(p) = \{(0, 0)\}, \quad \omega(p) = C_1 \quad \text{for } r \in (0, 1),$$

$$\alpha(p) = \omega(p) = C_1 \quad \text{for } r = 1,$$

$$\alpha(p) = C_1, \quad \omega(p) = C_2 \quad \text{for } r \in (1, 2),$$

$$\alpha(p) = \omega(p) = C_2 \quad \text{for } r = 2,$$

$$\alpha(p) = C_2, \quad \omega(p) = \varnothing \quad \text{for } r > 2$$

(see Fig. 3.2).

Fig. 3.2 The phase portrait
of Eq. (3.7)

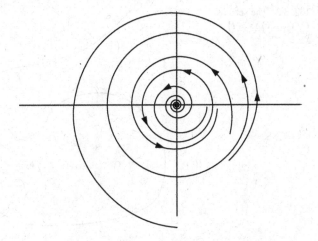

We also describe some properties of the α-limit set and the ω-limit set for flows and semiflows. With the exception of the connectedness of these sets, all the remaining properties are analogous to those already obtained for dynamical systems with discrete time.

Proposition 3.5 *Given a semiflow $\Phi = (\varphi_t)_{t \geq 0}$ of X, for each $x \in X$ the following properties hold:*

1. *$y \in \omega(x)$ if and only if there exists a sequence $t_k \nearrow +\infty$ in $\mathbb{R}^+$ such that $\varphi_{t_k}(x) \to y$ when $k \to \infty$;*
2. *if Φ is a topological semiflow, then $\omega(x)$ is forward Φ-invariant.*

Proof Both properties can be obtained repeating arguments in the proof of Proposition 3.1. □

Proposition 3.6 *Given a topological semiflow $\Phi = (\varphi_t)_{t \geq 0}$ of X, if the positive semiorbit $\gamma^+(x)$ of a point $x \in X$ has compact closure, then:*

1. *$\omega(x)$ is compact, connected and nonempty;*
2. *$\inf\{d(\varphi_t(x), y) : y \in \omega(x)\} \to 0$ when $t \to +\infty$.*

Proof With the exception of the connectedness of the ω-limit set, the remaining properties can be obtained repeating arguments in the proof of Proposition 3.2.

In order to show that $\omega(x)$ is connected, we proceed by contradiction. If $\omega(x)$ was not connected, then we could write it in the form $\omega(x) = A \cup B$ for some nonempty sets A and B such that

$$\overline{A} \cap B = A \cap \overline{B} = \emptyset.$$

Fig. 3.3 The set
$C \cap \{\varphi_s(x) : s > t\}$

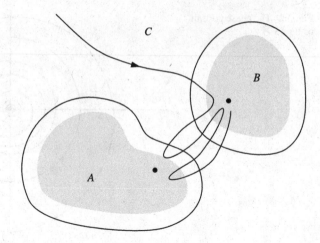

Since $\omega(x)$ is closed, we have

$$\overline{A} = \overline{A} \cap \omega(x) = \overline{A} \cap (A \cup B)$$
$$= (\overline{A} \cap A) \cup (\overline{A} \cap B) = A$$

and analogously $\overline{B} = B$. This shows that the sets A and B are also closed. This implies that they are at a positive distance, that is,

$$\delta := \inf\{d(a, b) : a \in A, \ b \in B\} > 0.$$

Now we consider the closed set

$$C = \{z \in X : d(z, y) \geq \delta/4 \text{ for } y \in \omega(x)\}. \tag{3.8}$$

We note that

$$C \cap \{\varphi_s(x) : s > t\} \neq \varnothing \tag{3.9}$$

for $t > 0$. Otherwise, the set $\{\varphi_s(x) : s > t\}$ would be completely contained in the $\delta/4$-neighborhood of A or in the $\delta/4$-neighborhood of B. Hence, by the first property in Proposition 3.5, we would have $\omega(x) \cap B = \varnothing$ or $\omega(x) \cap A = \varnothing$. But this is impossible since $\omega(x) = A \cup B$ with A and B nonempty. It follows from (3.9) that there exists a sequence $t_k \nearrow +\infty$ such that $\varphi_{t_k}(x) \in C$ for $k \in \mathbb{N}$ (see Fig. 3.3). Hence, it follows from the compactness of $C \cap \overline{\gamma^+(x)}$ and again from the first property in Proposition 3.5 that $C \cap \omega(x) \neq \varnothing$. On the other hand, it follows from (3.8) that $C \cap \omega(x) = \varnothing$. This contradiction shows that the set $\omega(x)$ is connected. $\quad\square$

In the case of flows, we have analogous results for the α-limit set.

Proposition 3.7 *Given a flow* $\Phi = (\varphi_t)_{t \in \mathbb{R}}$ *of* X, *for each* $x \in X$ *the following properties hold:*

1. $y \in \alpha(x)$ *if and only if there exists a sequence* $t_k \searrow -\infty$ *in* $\mathbb{R}^-$ *such that* $\varphi_{t_k}(x) \to y$ *when* $k \to \infty$;
2. *if* Φ *is a topological flow, then* $\alpha(x)$ *is backward* Φ-*invariant*.

Proposition 3.8 *Given a topological flow* $\Phi = (\varphi_t)_{t \in \mathbb{R}}$ *of* X, *if the negative semiorbit* $\gamma^-(x)$ *of a point* $x \in X$ *has compact closure, then*:

1. $\alpha(x)$ *is compact, connected and nonempty*;
2. $\inf\{d(\varphi_t(x), y) : y \in \alpha(x)\} \to 0$ *when* $t \to -\infty$.

3.3 Topological Recurrence

In this section we discuss some recurrence properties of the orbits of a topological dynamical system (with discrete time). Roughly speaking, a point x is recurrent if its orbit returns arbitrarily close to x.

3.3.1 Topological Transitivity

Let $f: X \to X$ be a continuous map.

Definition 3.5 A point $x \in X$ is said to be *(positively) recurrent* (with respect to f) if $x \in \omega(x)$.

It follows from Proposition 3.1 that a point x is recurrent if and only if there exists a sequence $n_k \nearrow \infty$ in $\mathbb{N}$ such that $f^{n_k}(x) \to x$ when $k \to \infty$. Moreover, the set of recurrent points (with respect to f) is forward invariant. Indeed, if $f^{n_k}(x) \to x$ with $n_k \nearrow \infty$ when $k \to \infty$, then also $f^{n_k+n}(x) \to f^n(x)$ when $k \to \infty$, for $n \in \mathbb{N}$.

For example, any periodic point x is recurrent since $x \in \gamma^+(x) = \omega(x)$.

Example 3.7 Consider the rotation $R_\alpha : S^1 \to S^1$. When α is rational, all points are periodic and thus, they are also recurrent. When α is irrational, for each $x \in S^1$, we have $\omega(x) = S^1$ and again all points are recurrent.

More generally, each point $x \in X$ with $\omega(x) = X$ is recurrent. Moreover, its positive semiorbit $\gamma^+(x)$ is dense in X (see Exercise 3.2).

Now we show that in compact metric spaces without isolated points, the existence of a dense positive semiorbit is equivalent to the following property.

Definition 3.6 A map $f: X \to X$ is called *topologically transitive* if given nonempty open sets $U, V \subset X$, there exists an $n \in \mathbb{N}$ such that $f^{-n}U \cap V \neq \varnothing$.

The following result establishes the desired equivalence.

Theorem 3.1 *Let* $f: X \to X$ *be a continuous map of a locally compact metric space with a countable basis. Then the following properties hold:*

1. *if* f *is topologically transitive, then there exists an* $x \in X$ *whose positive semiorbit* $\gamma^+(x)$ *is dense in* X;
2. *if* X *has no isolated points and there exists an* $x \in X$ *whose positive semiorbit* $\gamma^+(x)$ *is dense in* X, *then* f *is topologically transitive.*

Proof We first assume that f is topologically transitive. Given a nonempty open set $U \subset X$, the union $\bigcup_{n \in \mathbb{N}} f^{-n}U$ is dense in X since it intersects all open sets. Now let $\{U_i\}_{i \in \mathbb{N}}$ be a countable basis of X. Since any locally compact metric space is a Baire space (that is, it has the property that any countable intersection of dense open sets is dense), the set

$$Y = \bigcap_{i \in \mathbb{N}} \bigcup_{n \in \mathbb{N}} f^{-n}U_i$$

is nonempty. Given $x \in Y$, we have $x \in \bigcup_{n \in \mathbb{N}} f^{-n}U_i$ for $i \in \mathbb{N}$ and thus,

$$\gamma^+(x) \cap U_i \neq \varnothing \quad \text{for } i \in \mathbb{N}.$$

This shows that the positive semiorbit of x is dense in X.

Now we assume that X has no isolated points and that there exists an $x \in X$ with dense positive semiorbit. Let $U, V \subset X$ be nonempty open sets. Since X has no isolated points, the semiorbit $\gamma^+(x)$ visits infinitely often U and V. Hence, there exist $m, n \in \mathbb{N}$ with $m > n$ such that $f^m(x) \in U$ and $f^n(x) \in V$. Therefore,

$$x \in f^{-m}U \cap f^{-n}V = f^{-n}\big(f^{-(m-n)}U \cap V\big) \qquad (3.10)$$

and the set $f^{-(m-n)}U \cap V$ is nonempty. □

For example, it follows from Theorem 3.1 together with Example 3.2 that for each $\alpha \in \mathbb{R} \setminus \mathbb{Q}$ the rotation $R_\alpha: S^1 \to S^1$ is topologically transitive.

It is also common to use as an alternative definition of topological transitivity the existence of a dense positive semiorbit.

Finally, we show that for homeomorphisms of a compact metric space without isolated points, the existence of a dense orbit implies the existence of a dense positive semiorbit, possibly of another point.

Theorem 3.2 *Let* $f: X \to X$ *be a homeomorphism of a locally compact metric space with a countable basis and without isolated points. If there exists an* $x \in X$ *whose orbit* $\gamma(x)$ *is dense in* X, *then there exists a* $y \in X$ *whose positive semiorbit* $\gamma^+(y)$ *is dense in* X.

Proof A dense orbit $\gamma(x)$ visits infinitely often each open neighborhood of x (since x is not isolated). Thus, there exists a sequence n_k with $|n_k| \nearrow \infty$ such that

$f^{n_k}(x) \to x$ when $k \to \infty$. Since f is a homeomorphism, we also have

$$f^{n_k+m}(x) \to f^m(x) \quad \text{when } k \to \infty, \tag{3.11}$$

for each $m \in \mathbb{Z}$. We note that the sequence n_k takes infinitely many positive values or infinitely many negative values (or both). In the first case, it follows from (3.11) that the positive semiorbit $\gamma^+(x)$ is dense in X, which establishes the desired result. In the second case, the negative semiorbit $\gamma^-(x)$ is dense in X. Now let $U, V \subset X$ be nonempty open sets. Since $\gamma^-(x)$ is dense and X has no isolated points, there exist negative integers $m > n$ such that $f^m(x) \in U$ and $f^n(x) \in V$. Hence, property (3.10) holds and the set $f^{-(m-n)}U \cap V$ is nonempty. This shows that the map f is topologically transitive and it follows from Theorem 3.1 that there exists a dense positive semiorbit. $\qquad \square$

3.3.2 Topological Mixing

In this section we consider a recurrence property that is stronger than topological transitivity.

Definition 3.7 A map $f: X \to X$ is called *topologically mixing* if given nonempty open sets $U, V \subset X$, there exists an $n \in \mathbb{N}$ such that $f^{-m}U \cap V \neq \varnothing$ for $m \geq n$.

Clearly, any topologically mixing map is also topologically transitive. The following example shows that the converse is false.

Example 3.8 Let $R_\alpha: S^1 \to S^1$ be a rotation of the circle with $\alpha \in \mathbb{R} \setminus \mathbb{Q}$. Given $\varepsilon < 1/4$, we consider the open interval

$$U = (x - \varepsilon, x + \varepsilon) \subset S^1.$$

Since each preimage $R_\alpha^{-n}U$ is an open interval of length $2\varepsilon < 1/2$ and the orbit of x is dense, there exists a sequence $n_k \nearrow \infty$ in $\mathbb{N}$ such that $R_\alpha^{-n_k}(x) \to x + 1/2$ when $k \to \infty$. Thus, $R_\alpha^{-n_k}U \cap U = \varnothing$ for any sufficiently large k. This shows that the rotation R_α is not topologically mixing.

Example 3.9 Now we consider the expanding map $E_2: S^1 \to S^1$. By Example 3.4, there exists a point $x \in S^1$ whose positive semiorbit $\gamma^+(x)$ is dense in S^1. Hence, it follows from Theorem 3.1 that the map E_2 is topologically transitive.

Now we show that E_2 is also topologically mixing. Let $U, V \subset S^1$ be nonempty open sets and consider an open interval $I \subset V$ of the form

$$I = \left(0.x_1x_2 \cdots x_n, 0.x_1x_2 \cdots x_n 11 \cdots\right),$$

with the endpoints written in base-2. Given $y = 0.y_1 y_2 \cdots \in U$, the point

$$x = 0.x_1 x_2 \cdots x_n y_1 y_2 \cdots \in I$$

is in $E_2^{-n} U$ since $E_2^n(x) = y$. Therefore,

$$E_2^{-n} U \cap V \supset E_2^{-n} U \cap I \neq \varnothing.$$

This shows that the map E_2 is topologically mixing.

Example 3.10 Let $T_A \colon \mathbb{T}^2 \to \mathbb{T}^2$ be an automorphism of the torus $\mathbb{T}^2$ with $|\operatorname{tr} A| > 2$. By Exercise 2.12, we have $\det A = \pm 1$. We also have

$$\det(A - \lambda \mathrm{Id}) = \lambda^2 - \operatorname{tr} A \lambda + \det A,$$

and since $|\operatorname{tr} A| > 2$, the eigenvalues of the matrix A are the real numbers

$$\lambda_1 = \frac{\operatorname{tr} A + \sqrt{(\operatorname{tr} A)^2 - 4 \det A}}{2} \quad \text{and} \quad \lambda_2 = \frac{\operatorname{tr} A - \sqrt{(\operatorname{tr} A)^2 - 4 \det A}}{2}.$$

In particular, there exists a $\lambda > 1$ such that $\{|\lambda_1|, |\lambda_2|\} = \{\lambda, \lambda^{-1}\}$. Now we show that λ_1 and λ_2 are irrational. Clearly, λ_1 and λ_2 are rational if and only if $m^2 \pm 4 = l^2$ for some integer $l \in \mathbb{N}$, where $m = \operatorname{tr} A$. Hence,

$$(m - l)(m + l) = \pm 4$$

and thus,

$$m + l = 4 \quad \text{and} \quad m - l = 1$$

or

$$m + l = -1 \quad \text{and} \quad m - l = -4$$

(since $m + l > m - l$). It is easy to verify that these systems have no integer solutions. This implies that λ_1 and λ_2 are irrational. In particular, the eigendirections of A have irrational slopes.

Now let $U, V \subset \mathbb{T}^2$ be nonempty open sets and let $I \subset U$ be a line segment parallel to the eigendirection of A corresponding to the eigenvalue with modulus $\lambda^{-1} < 1$. Then $A^{-m} I \subset \mathbb{R}^2$ is a line segment of length $\lambda^m |I|$, where $|I|$ is the length of I. On the other hand, since the eigendirection of A corresponding to λ^{-1} has irrational slope, one can show that for any straight line $J \subset \mathbb{R}^2$ with this direction, the set $J/\mathbb{Z}^2$ is dense in $\mathbb{T}^2$. This implies that, given $\varepsilon > 0$, there exists an $L > 0$ such that for any line segment $J' \subset \mathbb{R}^2$ of length L with that direction, the set $J'/\mathbb{Z}^2$ is ε-dense in $\mathbb{T}^2$. In other words, the ε-neighborhood of $J'/\mathbb{Z}^2$ coincides with $\mathbb{T}^2$ (see Fig. 3.4). Now take $\varepsilon > 0$ such that V contains an open ball B of radius ε and take $n = n(\varepsilon) \in \mathbb{N}$ such that $\lambda^n |I| > L$ (recall that $\lambda > 1$). Since $\lambda^m |I| > L$ for $m \geq n$, we obtain

$$T_A^{-m} U \cap V \supset T_A^{-m} I \cap B \neq \varnothing$$

Fig. 3.4 An ε-dense line
segment in $\mathbb{T}^2$

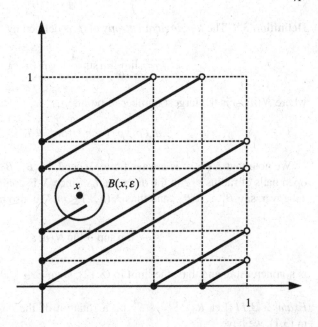

for $m \geq n$ (since $T_A^{-m} I$ is ε-dense in $\mathbb{T}^2$). This shows that the automorphism of the
torus T_A is topologically mixing.

3.4 Topological Entropy

In this section we introduce the notion of the topological entropy of a dynamical
system (with discrete time). Roughly speaking, topological entropy measures how
the orbits of a dynamical system move apart as time increases and thus, it can be
seen as a measure of the complexity of the dynamics. In addition to establishing
some basic properties of topological entropy, we also illustrate its computation with
several examples. The emphasis of this section is on the computation of topological
entropy. In particular, we describe several alternative characterizations of topological entropy that are particularly useful for its explicit computation. We also show
that topological entropy is a topological invariant, that is, it takes the same value for
topologically conjugate dynamical systems.

3.4.1 Basic Notions and Examples

Let $f\colon X \to X$ be a continuous map of a *compact* metric space X, say with distance d. For each $n \in \mathbb{N}$, we introduce a new distance on X by

$$d_n(x, y) = \max\bigl\{d\bigl(f^k(x), f^k(y)\bigr) : 0 \leq k \leq n - 1\bigr\}.$$

Definition 3.8 The *topological entropy* of f is defined by

$$h(f) = \lim_{\varepsilon \to 0} \limsup_{n \to \infty} \frac{1}{n} \log N(n, \varepsilon), \tag{3.12}$$

where $N(n, \varepsilon)$ is the largest number of points $p_1, \ldots, p_m \in X$ such that

$$d_n(p_i, p_j) \geq \varepsilon \quad \text{for } i \neq j.$$

We note that $N(n, \varepsilon)$ is always finite. Indeed, let $B_1, B_2, \ldots$ be a cover of X by open balls of radius $\varepsilon/2$ in the distance d_n. Since X is compact, there exists a finite subcover, say $B'_1, \ldots, B'_m$, and thus $N(n, \varepsilon) \leq m$. We also note that the function

$$\varepsilon \mapsto \limsup_{n \to \infty} \frac{1}{n} \log N(n, \varepsilon) \tag{3.13}$$

is nonincreasing and thus, the limit in (3.12) when $\varepsilon \to 0$ always exists.

Example 3.11 Let $R_\alpha \colon S^1 \to S^1$ be a rotation of the circle. For the distance d in (3.1), we have

$$d\big(R_\alpha(x), R_\alpha(y)\big) = d(x, y)$$

for $x, y \in S^1$. Thus, $d_n = d_1 = d$ for $n \in \mathbb{N}$ and

$$h(R_\alpha) = \lim_{\varepsilon \to 0} \limsup_{n \to \infty} \frac{1}{n} \log N(1, \varepsilon) = 0.$$

Example 3.12 Now we consider the expanding map $E_2 \colon S^1 \to S^1$. Since the function in (3.13) is nonincreasing, we have

$$h(E_2) = \lim_{k \to \infty} \limsup_{n \to \infty} \frac{1}{n} \log N(n, a_k)$$

for any sequence $(a_k)_{k \in \mathbb{N}} \subset \mathbb{R}^+$ such that $a_k \to 0$ when $k \to \infty$.

Now let us take $a_k = 2^{-(k+1)}$. We show that

$$N\big(n, 2^{-(k+1)}\big) = 2^{n+k} \quad \text{for } n, k \in \mathbb{N}. \tag{3.14}$$

We first observe that if $d(x, y) < 2^{-n}$, then

$$d_n(x, y) = d\big(E_2^{n-1}(x), E_2^{n-1}(y)\big) = 2^{n-1} d(x, y). \tag{3.15}$$

Now consider the points $p_i = i/2^{n+k}$ for $i = 0, \ldots, 2^{n+k} - 1$. It follows from (3.15) that

$$d_n(p_i, p_{i+1}) = 2^{-(k+1)} \quad \text{for } i = 0, \ldots, 2^{n+k} - 1.$$

Since there is no point p_j between p_i and p_{i+1}, we have

$$d_n(p_i, p_j) \geq 2^{-(k+1)} \quad \text{for } i \neq j$$

and thus,

$$N\left(n, 2^{-(k+1)}\right) \geq 2^{n+k}. \tag{3.16}$$

Now consider a set $A \subset S^1$ with cardinality at least $2^{n+k} + 1$. Clearly, there exist points $x, y \in A$ with $x \neq y$ such that $d(x, y) < 2^{-(n+k)}$. This implies that $d_n(x, y) < 2^{-(k+1)}$ and hence,

$$N\left(n, 2^{-(k+1)}\right) \leq 2^{n+k}. \tag{3.17}$$

It follows from (3.16) and (3.17) that property (3.14) holds. Thus,

$$h(E_2) = \lim_{k \to \infty} \limsup_{n \to \infty} \frac{1}{n} \log N\left(n, 2^{-(k+1)}\right)$$

$$= \lim_{k \to \infty} \limsup_{n \to \infty} \frac{n+k}{n} \log 2 = \log 2. \tag{3.18}$$

3.4.2 Topological Invariance

In this section we show that topological entropy is a topological invariant. We first introduce the notion of topological conjugacy.

Definition 3.9 Two maps $f : X \to X$ and $g : Y \to Y$, where X and Y are topological spaces, are said to be *topologically conjugate* if there exists a homeomorphism $H : X \to Y$ such that

$$H \circ f = g \circ H.$$

Then H is called a *topological conjugacy*.

Example 3.13 Consider the map $f : R \to R$ defined by $f(z) = z^2$ on the set

$$R = \left\{ z \in \mathbb{C} : |z| = 1 \right\}$$

(see Exercise 2.9). We also consider the continuous map $H : S^1 \to R$ defined by

$$H(x) = e^{2\pi i x}.$$

One can easily verify that H is a homeomorphism, with inverse given by

$$H^{-1}(z) = \frac{\arg z}{2\pi} \mod 1.$$

We have

$$(f \circ H)(x) = f\left(e^{2\pi i x}\right) = e^{4\pi i x}$$

and

$$(H \circ E_2)(x) = H(2x) = e^{4\pi i x}.$$

This shows that

$$H \circ E_2 = f \circ H$$

and thus, the maps E_2 and f are topologically conjugate.

We say that a certain quantity, such as, for example, topological entropy, is a *topological invariant* if it takes the same value for topologically conjugate dynamical systems. Now we show that topological entropy is a topological invariant.

Theorem 3.3 *Let $f: X \to X$ and $g: Y \to Y$ be continuous maps of compact metric spaces. If f and g are topologically conjugate, then $h(f) = h(g)$.*

Proof Let $H: X \to Y$ be a homeomorphism such that

$$H \circ f = g \circ H. \tag{3.19}$$

Since the map H is uniformly continuous, given $\varepsilon > 0$, there exists a $\delta > 0$ such that

$$d_Y\left(H(x), H(y)\right) < \varepsilon \quad \text{when } d_X(x, y) < \delta, \tag{3.20}$$

where d_X and d_Y are, respectively, the distances on X and Y. We note that $\delta \to 0$ when $\varepsilon \to 0$. On the other hand, it follows from (3.19) that

$$H\left(f^m(x)\right) = g^m(H(x))$$

for $m \in \mathbb{N}$ and $x \in X$. Hence, by (3.20), if $p_1, \ldots, p_m \in Y$ are such that

$$\max\{d_Y\left(g^m(q_i), g^m(q_j)\right) : m = 0, \ldots, n-1\} \geq \varepsilon \quad \text{for } i \neq j,$$

where $q_i = H(p_i)$, then

$$\max\{d_X\left(f^m(p_i), f^m(p_j)\right) : m = 0, \ldots, n-1\} \geq \delta \quad \text{for } i \neq j.$$

This shows that

$$N_f(n, \delta) \geq N_g(n, \varepsilon), \tag{3.21}$$

where we indicated the particular dynamical system used in Definition 3.8. It follows from (3.21) that

$$\limsup_{n \to \infty} \frac{1}{n} \log N_f(n, \delta) \geq \limsup_{n \to \infty} \frac{1}{n} \log N_g(n, \varepsilon)$$

for each $\varepsilon > 0$. Letting $\varepsilon \to 0$, we have $\delta \to 0$ and thus $h(f) \geq h(g)$. Now we rewrite identity (3.19) in the form

$$H^{-1} \circ g = f \circ H^{-1}.$$

Repeating the previous argument with H replaced by H^{-1}, we obtain $h(g) \geq h(f)$. Therefore, $h(f) = h(g)$. $\square$

Example 3.14 The map f in Example 3.13 is topologically conjugate to the expanding map E_2. Hence, it follows from Theorem 3.3 together with Example 3.12 (see (3.18)) that

$$h(f) = h(E_2) = \log 2.$$

3.4.3 Alternative Characterizations

In this section we describe several alternative characterizations of topological entropy. These are particularly useful in the computation of the entropy.

Definition 3.10 Given $n \in \mathbb{N}$ and $\varepsilon > 0$, we denote by $M(n, \varepsilon)$ the least number of points $p_1, \ldots, p_m \in X$ such that each $x \in X$ satisfies $d_n(x, p_i) < \varepsilon$ for some i.

Definition 3.11 Given $n \in \mathbb{N}$ and $\varepsilon > 0$, we denote by $C(n, \varepsilon)$ the least number of elements of a cover of X by sets $U_1, \ldots, U_m$ with

$$\sup\{d_n(x, y) : x, y \in U_i\} < \varepsilon \quad \text{for } i = 1, \ldots, m. \tag{3.22}$$

The supremum in (3.22) is called the d_n-*diameter of* U_i.

We have the following relations between these numbers and $N(n, \varepsilon)$ (see Definition 3.8).

Proposition 3.9 *For each* $n \in \mathbb{N}$ *and* $\varepsilon > 0$, *we have*

$$C(n, 2\varepsilon) \leq M(n, \varepsilon) \leq N(n, \varepsilon) \leq M(n, \varepsilon/2) \leq C(n, \varepsilon/2). \tag{3.23}$$

Proof We establish successively each of the inequalities:

1. Take points $p_1, \ldots, p_m \in X$ such that each $x \in X$ satisfies $d_n(x, p_i) < \varepsilon$ for some i, where $m = M(n, \varepsilon)$. Clearly, the d_n-open balls

$$B_n(p_i, \varepsilon) = \{x \in X : d_n(x, p_i) < \varepsilon\}$$

cover X. Since $B_n(p_i, \varepsilon)$ has d_n-diameter 2ε, we conclude that $m \geq C(n, 2\varepsilon)$.

2. Now let $p_1, \ldots, p_m \in X$ be points such that $d_n(p_i, p_j) \geq \varepsilon$ for $i \neq j$, where $m = N(n, \varepsilon)$. We note that each $x \in X \setminus \{p_1, \ldots, p_m\}$ satisfies $d_n(x, p_i) < \varepsilon$ for some i. Hence, $M(n, \varepsilon) \leq m$.

3. For the third inequality, we note that no d_n-open ball of radius $\varepsilon/2$ contains two points at a d_n-distance ε. Thus $N(n, \varepsilon) \le M(n, \varepsilon/2)$.
4. Finally, let $U_1, \ldots, U_m$ be a cover of X by sets of d_n-diameter less than $\varepsilon/2$, where $m = C(n, \varepsilon/2)$. Now take a point $p_i \in U_i$ for each i. Clearly, $B_n(p_i, \varepsilon/2) \supset U_i$ and these d_n-balls form a cover of X. Hence, $M(n, \varepsilon/2) \le C(n, \varepsilon/2)$.

This completes the proof of the proposition.   □

Now we obtain several alternative formulas for the topological entropy of a dynamical system.

Theorem 3.4 *If $f : X \to X$ is a continuous map of a compact metric space, then*

$$h(f) = \lim_{\varepsilon \to 0} \liminf_{n \to \infty} \frac{1}{n} \log N(n, \varepsilon)$$

$$= \lim_{\varepsilon \to 0} \limsup_{n \to \infty} \frac{1}{n} \log M(n, \varepsilon)$$

$$= \lim_{\varepsilon \to 0} \liminf_{n \to \infty} \frac{1}{n} \log M(n, \varepsilon)$$

$$= \lim_{\varepsilon \to 0} \lim_{n \to \infty} \frac{1}{n} \log C(n, \varepsilon). \tag{3.24}$$

Proof We first establish the existence of the limit when $n \to \infty$ in the last expression in (3.24).

Lemma 3.1 *Given $m, n \in \mathbb{N}$ and $\varepsilon > 0$, we have*

$$C(m + n, \varepsilon) \le C(m, \varepsilon)C(n, \varepsilon).$$

Proof Let $U_1, \ldots, U_k$ be a cover of X by sets of d_n-diameter less than ε, where $k = C(n, \varepsilon)$. Let also $V_1, \ldots, V_l$ be a cover of X by sets of d_m-diameter less than ε, where $l = C(m, \varepsilon)$. Then the sets $U_i \cap f^{-n} V_j$, with $i = 1, \ldots, k$ and $j = 1, \ldots, l$, form a cover of X and have d_{m+n}-diameter less than ε since

$$d_{m+n}(x, y) = \max\{d_n(x, y), d_m\left(f^n(x), f^n(y)\right)\}.$$

Thus,

$$C(m + n, \varepsilon) \le lk = C(m, \varepsilon)C(n, \varepsilon),$$

which yields the desired inequality.   □

Now we establish an auxiliary result.

Lemma 3.2 *If $(c_n)_{n \in \mathbb{N}}$ is a sequence of real numbers such that*

$$c_{m+n} \leq c_m + c_n \tag{3.25}$$

for $m, n \in \mathbb{N}$, then the limit

$$\lim_{n \to \infty} \frac{c_n}{n} = \inf\left\{ \frac{c_n}{n} : n \in \mathbb{N} \right\}$$

exists.

Proof Given integers $n, k \in \mathbb{N}$, write $n = qk + r$ with $q \in \mathbb{N} \cup \{0\}$ and $r \in \{0, \ldots, q - 1\}$. We have

$$\frac{c_n}{n} \leq \frac{c_{qk} + c_r}{qk + r} \leq \frac{q c_k + c_r}{qk + r}$$

and thus,

$$\limsup_{n \to \infty} \frac{c_n}{n} \leq \frac{c_k}{k}$$

since $q \to \infty$ when $n \to \infty$ (for a fixed k). Since k is arbitrary, this implies that

$$\limsup_{n \to \infty} \frac{c_n}{n} \leq \inf\left\{ \frac{c_k}{k} : k \in \mathbb{N} \right\} \leq \liminf_{n \to \infty} \frac{c_n}{n},$$

which yields the desired result. □

It follows from Lemmas 3.1 and 3.2 that the limit

$$\lim_{n \to \infty} \frac{1}{n} \log C(n, \varepsilon) = \inf\left\{ \frac{1}{n} \log C(n, \varepsilon) : n \in \mathbb{N} \right\}$$

exists. Using (3.23), we obtain

$$\lim_{n \to \infty} \frac{1}{n} \log C(n, 2\varepsilon) \leq \liminf_{n \to \infty} \frac{1}{n} \log M(n, \varepsilon)$$

$$\leq \liminf_{n \to \infty} \frac{1}{n} \log N(n, \varepsilon)$$

$$\leq \limsup_{n \to \infty} \frac{1}{n} \log N(n, \varepsilon)$$

$$\leq \limsup_{n \to \infty} \frac{1}{n} \log M(n, \varepsilon/2)$$

$$\leq \lim_{n \to \infty} \frac{1}{n} \log C(n, \varepsilon/2)$$

and letting $\varepsilon \to 0$ yields the inequalities

$$\lim_{\varepsilon \to 0} \lim_{n \to \infty} \frac{1}{n} \log C(n, 2\varepsilon) \le \lim_{\varepsilon \to 0} \liminf_{n \to \infty} \frac{1}{n} \log M(n, \varepsilon)$$

$$\le \lim_{\varepsilon \to 0} \liminf_{n \to \infty} \frac{1}{n} \log N(n, \varepsilon)$$

$$\le h(f)$$

$$\le \lim_{\varepsilon \to 0} \limsup_{n \to \infty} \frac{1}{n} \log M(n, \varepsilon/2)$$

$$\le \lim_{\varepsilon \to 0} \lim_{n \to \infty} \frac{1}{n} \log C(n, \varepsilon/2).$$

The equality of the first and the last terms establishes the desired result.   □

Now we use Theorem 3.4 to compute the topological entropy of a class of auto-morphisms of the torus $\mathbb{T}^2$.

Example 3.15 Let $T_A \colon \mathbb{T}^2 \to \mathbb{T}^2$ be an automorphism of the torus as in Example 3.10. We recall that along the eigendirections of A the distances are multiplied by λ or λ^{-1}, for some $\lambda > 1$.

Now we consider a cover of $\mathbb{T}^2$ by d_n-open balls $B_n(p_i, \varepsilon)$. We have

$$B_n(p_i, \varepsilon) = \bigcap_{k=0}^{n-1} T_A^{-k} B\big(T_A^k(p_i), \varepsilon\big)$$

and thus, there exists a $C > 0$ (independent of n, ε and i) such that the area of the ball $B_n(p_i, \varepsilon)$ is at most $C\lambda^{-n}\varepsilon^2$. Hence,

$$M(n, \varepsilon) \ge \frac{1}{C\lambda^{-n}\varepsilon^2}$$

and it follows from Theorem 3.4 that

$$h(f) = \lim_{\varepsilon \to 0} \liminf_{n \to \infty} \frac{1}{n} \log M(n, \varepsilon) \ge \log \lambda. \tag{3.26}$$

We also consider partitions of $\mathbb{T}^2$ by parallelograms with sides parallel to the eigendirections of A (see Fig. 3.5). More precisely, we consider a partition of $\mathbb{T}^2$ by parallelograms P_i with sides of length $\varepsilon\lambda^{-n}$ and ε, up to a multiplicative constant, along the eigendirections of λ and λ^{-1}, respectively. Now we note that there exists a $D > 1$ (independent of n, ε and i) such that each P_i has area at least $D^{-1}\lambda^{-n}\varepsilon^2$ and has d_n-diameter less than $D\varepsilon$. Thus,

$$C(n, D\varepsilon) \le \frac{1}{D^{-1}\lambda^{-n}\varepsilon^2}$$

Fig. 3.5 A partition of the torus $\mathbb{T}^2$ into parallelograms

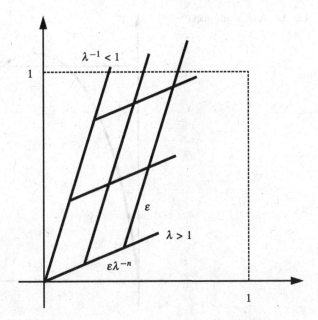

and by Theorem 3.4, we have

$$h(f) = \lim_{\varepsilon \to 0} \lim_{n \to \infty} \frac{1}{n} \log C(n, \varepsilon) \leq \log \lambda.$$

Together with (3.26) this shows that $h(f) = \log \lambda$.

3.4.4 Expansive Maps

In this section we describe a class of maps for which the limit when $\varepsilon \to 0$ in the definition of topological entropy is not necessary.

Definition 3.12 A map $f : X \to X$ is called *(positively) expansive* if there exists a $\delta > 0$ such that if

$$d(f^n(x), f^n(y)) < \delta \quad \text{for all } n \geq 0,$$

then $x = y$.

Example 3.16 The expanding map $E_m : S^1 \to S^1$ is expansive. Indeed, if $d(x, y) < 1/m^2$ and $x \neq y$, then there exists an $n \in \mathbb{N}$ such that

$$d(E_m^n(x), E_m^n(y)) = m^n d(x, y) \geq \frac{1}{m^2}.$$

Fig. 3.6 A quadratic map

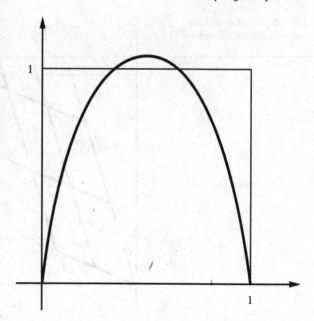

This implies that if

$$d\big(E_m^n(x), E_m^n(y)\big) < \frac{1}{m^2} \quad \text{for all } n \geq 0,$$

then $x = y$ and the expanding map E_m is expansive.

Example 3.17 Given $a > 4$, let $f : [0, 1] \to \mathbb{R}$ be the quadratic map

$$f(x) = ax(1 - x)$$

(see Fig. 3.6). The set

$$X = \bigcap_{n=0}^{\infty} f^{-n}[0, 1] \tag{3.27}$$

is compact and forward f-invariant. In particular, one can consider the restriction

$$f|X : X \to X.$$

Since $f(x) = 1$ for $x = (1 \pm c)/2$, where $c = \sqrt{1 - 4/a}$, we have

$$|f'(x)| = a|1 - 2x| \geq ac \quad \text{for } x \in X. \tag{3.28}$$

Now let us assume that $a > 4$ is so large that $ac > 1$, or equivalently that $a > 2 + \sqrt{5}$. Given $x, y \in X$ such that

$$\big|f^k(x) - f^k(y)\big| < c \quad \text{for } k \in \mathbb{N} \cup \{0\},$$

we have

$$f^k(x), f^k(y) \in I_1 \quad \text{or} \quad f^k(x), f^k(y) \in I_2,$$

where

$$I_1 = [0, (1-c)/2] \quad \text{and} \quad I_2 = [(1+c)/2, 1].$$

It then follows from (3.28) that

$$c > |f^k(x) - f^k(y)| \geq (ac)^k |x - y| \quad \text{for } k \in \mathbb{N}.$$

Since $ac > 1$, we conclude that $x = y$ and the map $f|X$ is expansive.

Now we consider the particular case of the expansive maps in the definition of topological entropy and we show that the limit when $\varepsilon \to 0$ is not necessary in any of the formulas in (3.24), provided that ε is sufficiently small.

Theorem 3.5 *Let $f: X \to X$ be a continuous expansive map of a compact metric space. Then*

$$h(f) = \lim_{n \to \infty} \frac{1}{n} \log N(n, \alpha)$$

$$= \lim_{n \to \infty} \frac{1}{n} \log M(n, \alpha)$$

$$= \lim_{n \to \infty} \frac{1}{n} \log C(n, \alpha) \qquad (3.29)$$

for any sufficiently small $\alpha > 0$.

Proof Take constants $\varepsilon, \alpha > 0$ such that $0 < \varepsilon < \alpha < \delta$, where δ is the constant in Definition 3.12. Now let $A \subset X$ be a set with $\operatorname{card} A = N(n, \varepsilon)$ such that $d_n(x, y) \geq \varepsilon$ for any $x, y \in A$ with $x \neq y$. We show that there exists an $m = m(\varepsilon, \alpha) \in \mathbb{N}$ such that if $d(x, y) \geq \varepsilon$, then

$$d(f^i(x), f^i(y)) > \alpha \quad \text{for some } i \in \{0, \ldots, m\}. \qquad (3.30)$$

Given

$$q \in K := \{(x, y) \in X \times X : d(x, y) \geq \varepsilon\},$$

there exist an open ball $B(q) \subset X \times X$ centered at q and an integer $i = i(q) \in \mathbb{N} \cup \{0\}$ such that if $(x, y) \in B(q)$, then

$$d(f^i(x), f^i(y)) > \delta > \alpha$$

(recall that f is continuous and expansive). The balls $B(q)$ cover the compact set K and hence, there exists a finite subcover $B(q_j)$, with $j = 1, \ldots, p$. Taking

$$m = \max\{i(q_j) : j = 1, \ldots, p\},$$

we obtain property (3.30) for $(x, y) \in K$. This implies that when $d_n(x, y) \geq \varepsilon$ and hence, for $x, y \in A$ with $x \neq y$, we have

$$d_n\big(f^j(x), f^j(y)\big) > \alpha \quad \text{for some } j \in \{0, \ldots, m\}.$$

Thus, for $z, w \in f^{-m}A$ with $f^m(z) \neq f^m(w)$, we have

$$d_{n+2m}(z, w) \geq \max\big\{d_n\big(f^i(z), f^i(w)\big) : i = m, \ldots, 2m\big\}$$

$$= \max\big\{d_n\big(f^{j+m}(z), f^{j+m}(w)\big) : j = 0, \ldots, m\big\} > \alpha$$

since $f^m(z), f^m(w) \in A$. This yields the inequality

$$N(n + 2m, \alpha) \geq N(n, \varepsilon).$$

It follows from Proposition 3.9 that

$$N(n, \varepsilon) \leq N(n + 2m, \alpha)$$

$$\leq M(n + 2m, \alpha/2)$$

$$\leq C(n + 2m, \alpha/2)$$

$$\leq C(n + 2m, \varepsilon/2).$$

Thus, applying Theorem 3.4, we conclude that

$$\limsup_{n \to \infty} \frac{1}{n} \log N(n, \varepsilon) \leq \limsup_{n \to \infty} \frac{1}{n} \log N(n + 2m, \alpha)$$

$$\leq \limsup_{n \to \infty} \frac{1}{n} \log M(n + 2m, \alpha/2)$$

$$\leq \lim_{n \to \infty} \frac{1}{n} \log C(n + 2m, \alpha/2)$$

$$\leq \lim_{n \to \infty} \frac{1}{n} \log C(n + 2m, \varepsilon/2). \tag{3.31}$$

Letting $\varepsilon \to 0$ yields the inequalities

$$h(f) \leq \limsup_{n \to \infty} \frac{1}{n} \log N(n, \alpha)$$

$$\leq \limsup_{n \to \infty} \frac{1}{n} \log M(n, \alpha/2)$$

$$\leq \lim_{n \to \infty} \frac{1}{n} \log C(n, \alpha/2) \leq h(f). \tag{3.32}$$

One can also replace each $\limsup$ in (3.31) by $\liminf$ and letting $\varepsilon \to 0$, we obtain

$$h(f) \leq \liminf_{n \to \infty} \frac{1}{n} \log N(n, \alpha)$$

$$\leq \liminf_{n \to \infty} \frac{1}{n} \log M(n, \alpha/2)$$

$$\leq \lim_{n \to \infty} \frac{1}{n} \log C(n, \alpha/2) \leq h(f). \tag{3.33}$$

The identities in (3.29) now follow readily from (3.32) and (3.33).   □

By Example 3.16, the expanding maps E_m are expansive and thus, their topological entropies are given by (3.29). In particular, this implies that the limits in (3.18) when $k \to \infty$ are not necessary.

Now we consider another expansive map.

Example 3.18 Consider the restriction $E_4|A \colon A \to A$, where A is the compact forward E_4-invariant set in (2.11). We proceed in an analogous manner to that in Example 3.12. We first note that if $d(x, y) < 4^{-n}$, then

$$d_n(x, y) = d\left(E_4^{n-1}(x), E_4^{n-1}(y)\right) = 4^{n-1} d(x, y). \tag{3.34}$$

Given $k \in \mathbb{N}$, consider the 2^{n+k+1} points x_i on the boundary of the set

$$\bigcap_{m=0}^{n+k-1} E_4^{-m}\left([0, 1/4] \cup [2/4, 3/4]\right).$$

It follows from (3.34) that

$$d_n(x_i, x_j) \geq 4^{n-1} \cdot \frac{1}{4^{n+k}} = 4^{-(k+1)}$$

for $i \neq j$ and thus,

$$N\left(n, 4^{-(k+1)}\right) \geq 2^{n+k+1}.$$

On the other hand, given a set $B \subset A$ with at least $2^{n+k+1} + 1$ points, there exist $x, y \in B$ with $x \neq y$ such that $d(x, y) < 4^{-(n+k)}$ and thus $d_n(x, y) < 4^{-(k+1)}$. This implies that

$$N\left(n, 4^{-(k+1)}\right) = 2^{n+k+1} \quad \text{for } n, k \in \mathbb{N}.$$

Since E_4 is an expansive map (by Example 3.16), the same happens to the restriction $E_4|A$. It then follows from Theorem 3.5 that

$$h(E_4|A) = \lim_{n \to \infty} \frac{1}{n} \log N\left(n, 4^{-(k+1)}\right)$$

$$= \lim_{n \to \infty} \frac{n+k+1}{n} \log 2 = \log 2.$$

3.5 Exercises

Exercise 3.1 Show that if $f: X \to X$ is a homeomorphism, then for each $x \in X$ the sets $\alpha(x)$ and $\omega(x)$ are f-invariant.

Exercise 3.2 Given a map $f: \mathbb{R}^n \to \mathbb{R}^n$, show that the positive semiorbit $\gamma^+(x)$ is dense if and only if $\omega(x) = \mathbb{R}^n$.

Exercise 3.3 Determine whether there exists a differential equation in $\mathbb{R}^2$ whose flow:

1. has an ω-limit set that is the boundary of a square;
2. has a disconnected ω-limit set.

Exercise 3.4 Sketch the phase portrait of a differential equation in $\mathbb{R}^2$ whose flow has an ω-limit set that is the boundary of a triangle.

Exercise 3.5 Let φ_t be a flow determined by a differential equation $x' = f(x)$ for some C^1 function $f: \mathbb{R}^2 \to \mathbb{R}^2$. Show that if $L \subset \mathbb{R}^2$ is a *transversal* to f (that is, a line segment such that for each $x \in L$ the directions of L and $f(x)$ generate $\mathbb{R}^2$), then for each $x \in \mathbb{R}^2$ the set $\omega(x) \cap L$ contains at most one point.

Exercise 3.6 Show that no increasing homeomorphism $f: I \to I$, where $I \subset \mathbb{R}$ is an interval, is topologically transitive.

Exercise 3.7 Let $f: I \to I$ be a continuous onto map, where $I \subset \mathbb{R}$ is an interval. Show that the following properties are equivalent:

1. f is topologically transitive;
2. for any open interval $J \subset I$, the set $\bigcup_{n=0}^{\infty} f^{-n} J$ is dense in I;
3. for any open interval $J \subset I$, the set $\bigcup_{n=0}^{\infty} f^n(J)$ is dense in I.

Exercise 3.8 Show that:

1. for each $\alpha \in \mathbb{Q}$, the rotation $R_\alpha: S^1 \to S^1$ is not topologically mixing;
2. the expanding map E_m is topologically mixing.

Exercise 3.9 Determine whether the maps $f, g: \mathbb{R} \to \mathbb{R}$ are topologically conjugate for:

1. $f(x) = x$ and $g(x) = x^2$;
2. $f(x) = x/3$ and $g(x) = 2x$;
3. $f(x) = 2x$ and $g(x) = x^3$.

Exercise 3.10 Given an integer $m > 1$, consider the map $f: R \to R$ defined by $f(z) = z^m$ on the set

$$R = \{z \in \mathbb{C} : |z| = 1\}.$$

Show that E_m and f are topologically conjugate.

Exercise 3.11 Compute the topological entropy of the map $f : \mathbb{T}^n \to \mathbb{T}^n$ defined by $f(x) = x + v$, where $v \in \mathbb{R}^n$.

Exercise 3.12 Show that $h(E_m) = \log m$ for each integer $m > 1$.

Exercise 3.13 Show that if $f : X \to X$ is a homeomorphism of a compact metric space, then $h(f^{-1}) = h(f)$.

Exercise 3.14 Let $f : X \to X$ be a continuous map of a compact metric space. Show that if $X = \bigcup_{i=1}^m X_i$, where each set X_i is closed and forward f-invariant, then

$$h(f) = \max\{h(f|X_i) : i = 1, \ldots, m\}.$$

Exercise 3.15 Show that if $T_A : \mathbb{T}^n \to \mathbb{T}^n$ is an automorphism of the torus induced by a matrix A without eigenvalues with modulus 1, then

$$h(T_A) = \sum_{i=1}^n \max\{0, \log|\lambda_i|\},$$

where $\lambda_1, \ldots, \lambda_n$ are the eigenvalues of A, counted with their multiplicities.

Exercise 3.16 Compute the topological entropy of the endomorphism of the torus $\mathbb{T}^2$ induced by the matrix

$$A = \begin{pmatrix} 2 & 0 \\ 0 & 3 \end{pmatrix}.$$

Exercise 3.17 Show that if $f : X \to X$ and $g : Y \to Y$ are continuous maps of compact metric spaces (X, d_X) and (Y, d_Y), then

$$h(v) = h(f) + h(g)$$

for the map $v : X \times Y \to X \times Y$ defined by

$$v(x, y) = \big(f(x), g(y)\big),$$

with the distance on $X \times Y$ given by

$$d\big((x, y), (x', y')\big) = \max\{d_X(x, x'), d_Y(y, y')\}.$$

Exercise 3.18 Let $f : X \to X$ be a continuous map of a compact metric space and take $k \in \mathbb{N}$.

1. Writing $d_{n,f} = d_n$ and $N_f(n, \varepsilon) = N(n, \varepsilon)$, show that $d_{n,f^k}(x, y) \le d_{nk,f}(x, y)$ and thus,

$$N_{f^k}(n, \varepsilon) \le N_f(nk, \varepsilon).$$

2. Conclude that $h(f^k) \leq kh(f)$.
3. Show that

$$\lim_{\varepsilon \to 0} \limsup_{n \to \infty} \frac{1}{n} \log N_f(nk, \varepsilon) \leq h(f^k).$$

Hint: By the uniform continuity of f, given $\varepsilon > 0$, there exists a $\delta(\varepsilon) \in (0, \varepsilon)$ such that $d_k(x, y) < \varepsilon$ when $d(x, y) < \delta(\varepsilon)$. Hence, it follows from

$$d_{nk, f}(x, y) = \max \{ d_k(f^{ik}(x), f^{ik}(y)) : 0 \leq i \leq n - 1 \}$$

that

$$d_{n, f^k}(x, y) \geq \delta(\varepsilon) \quad \text{when } d_{nk, f}(x, y) \geq \varepsilon,$$

which yields the inequality

$$N_f(nk, \varepsilon) \leq N_{f^k}(n, \delta(\varepsilon)).$$

4. Use inequality (3.23) and Theorem 3.4 to conclude that $h(f^k) \geq kh(f)$.

The exercise shows that $h(f^k) = kh(f)$.

Exercise 3.19 Show that if $f: X \to X$ is a continuous map of a compact metric space and $h(f^n) \leq an + b$ for any $n \in \mathbb{N}$, then $h(f) \leq a$.

Exercise 3.20 Let $f, g: X \to X$ be continuous maps of a compact metric space. Show that if

$$\left| h(f^n) - h(g^n) \right| < \sqrt{n}$$

for any $n \in \mathbb{N}$, then $h(f) = h(g)$.

Chapter 4
Low-Dimensional Dynamics

In this chapter we consider several classes of dynamical systems in low-dimensional spaces. This essentially means dimension 1 for maps and dimension 2 for flows. In particular, we consider homeomorphisms and diffeomorphisms of the circle, continuous maps of a compact interval and flows defined by autonomous differential equations in the plane. For the orientation-preserving homeomorphisms of the circle, we consider the notion of rotation number and we describe the behavior of the orbits depending on whether it is rational or irrational. We also show that any sufficiently regular orientation-preserving diffeomorphism of the circle with irrational rotation number is topologically conjugate to a rotation. For the continuous maps of an interval, we study the existence of periodic points and we establish Sharkovsky's theorem relating the existence of periodic points with different periods. Finally, we establish the Poincaré–Bendixson theorem for differential equations in the plane.

4.1 Homeomorphisms of the Circle

In this section we consider orientation-preserving homeomorphisms of the circle and we introduce the notion of rotation number. Essentially, it gives the average angular speed with which a point of the circle rotates (or translates) under the action of the homeomorphism.

4.1.1 Lifts

In order to introduce the notion of a lift, we consider the projection $\pi: \mathbb{R} \to S^1$ defined by

$$\pi(x) = [x].$$

We often represent the equivalence class $[x]$ by its representative in the interval $[0, 1)$, that is, by the number $x - \lfloor x \rfloor$, where $\lfloor x \rfloor$ is the integer part of x.

L. Barreira, C. Valls, *Dynamical Systems*, Universitext, DOI 10.1007/978-1-4471-4835-7_4,

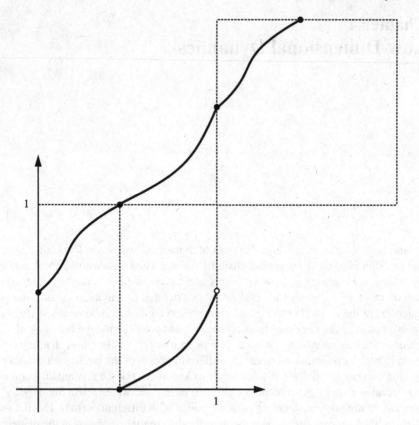

Fig. 4.1 The construction of a lift

Now let $f \colon S^1 \to S^1$ be a homeomorphism of the circle.

Definition 4.1 A continuous function $F \colon \mathbb{R} \to \mathbb{R}$ is said to be a *lift* of f if

$$f \circ \pi = \pi \circ F \qquad\qquad (4.1)$$

(see Fig. 4.1).

Example 4.1 Given $\alpha \in \mathbb{R}$, consider the rotation $R_\alpha \colon S^1 \to S^1$ given by

$$R_\alpha(x) = x + \alpha \bmod 1. \qquad\qquad (4.2)$$

Clearly, R_α is a homeomorphism. Given $k \in \mathbb{Z}$, the function $F \colon \mathbb{R} \to \mathbb{R}$ defined by

$$F(x) = x + \alpha + k \qquad\qquad (4.3)$$

satisfies

$$\pi(F(x)) = \pi(x + \alpha + k)$$
$$= x + \alpha + k \bmod 1$$
$$= \pi(x) + \alpha \bmod 1$$
$$= R_\alpha(\pi(x)).$$

Hence, F is a lift of R_α.

Example 4.2 Given $\beta \in \mathbb{R}$, consider the continuous function $f \colon S^1 \to S^1$ defined by

$$f(x) = x + \beta \sin(2\pi x) \bmod 1. \tag{4.4}$$

We first show that f is a homeomorphism for $|\beta| < 1/(2\pi)$. The function $F \colon \mathbb{R} \to \mathbb{R}$ defined by

$$F(x) = x + \beta \sin(2\pi x) \tag{4.5}$$

is increasing since

$$F'(x) = 1 + 2\pi\beta\cos(2\pi x) \geq 1 - 2\pi|\beta| > 0.$$

In particular, for $x \in [0, 1)$, we have

$$F(x) < F(1) = 1 \tag{4.6}$$

and thus, the function f is one-to-one and onto. Since f is continuous, it maps compact sets to compact sets. Thus, it also maps open sets to open sets, which shows that its inverse is continuous. Hence, it is a homeomorphism. Moreover, it follows from (4.6) that

$$\pi(F(x)) = x + \beta\sin(2\pi x) \bmod 1$$
$$= x - \lfloor x \rfloor + \beta\sin(2\pi x)$$
$$= x - \lfloor x \rfloor + \beta\sin\big(2\pi(x - \lfloor x \rfloor)\big)$$
$$= f(\pi(x))$$

and F is a lift of f.

The lifts of a homeomorphism have the following properties.

Proposition 4.1 *Let $f \colon S^1 \to S^1$ be a homeomorphism. Then:*

1. *f has lifts;*
2. *if F and G are lifts of f, then there exists a $k \in \mathbb{Z}$ such that $G - F = k$;*
3. *any lift of f is a homeomorphism of $\mathbb{R}$.*

Proof We define a function $F: \mathbb{R} \to \mathbb{R}$ by

$$F(x) = f(x - \lfloor x \rfloor) + \lfloor x \rfloor, \tag{4.7}$$

where $f(x - \lfloor x \rfloor)$ is the representative in the interval $[0, 1)$. Since $x - \lfloor x \rfloor$ and $\lfloor x \rfloor$ are continuous functions on $\mathbb{R} \setminus \mathbb{Z}$, so too is F. Moreover, for each $k \in \mathbb{Z}$, we have

$$F(k) = f(0) + k, \quad F(k^-) = f(1) + k \quad \text{and} \quad F(k^+) = f(0) + k.$$

Since the function f takes values in S^1, we have $f(0) = f(1)$ and thus,

$$F(k) = F(k^-) = F(k^+)$$

for $k \in \mathbb{Z}$. This shows that the function F is continuous on $\mathbb{R}$. We also have

$$\pi(F(x)) = f(x - \lfloor x \rfloor) = f(\pi(x))$$

and hence, F is a lift of f.

Now let F and G be lifts of f. Then

$$\pi \circ F = \pi \circ G = f \circ \pi. \tag{4.8}$$

It follows from the first identity in (4.8) that for each $x \in \mathbb{R}$, there exists a $p(x) \in \mathbb{Z}$ such that

$$G(x) - F(x) = p(x).$$

Since F and G are continuous, the function $x \mapsto p(x)$ is also continuous. Moreover, since it takes only integer values, it must be constant and thus, there exists a $k \in \mathbb{Z}$ such that

$$G(x) - F(x) = p(x) = k$$

for any $x \in \mathbb{R}$.

For the last property, since the lifts are unique up to an additive constant (by the second property), it is sufficient to show that the lift F constructed in (4.7) is a homeomorphism. Consider the continuous function $H: \mathbb{R} \to \mathbb{R}$ defined by

$$H(x) = f^{-1}(x - \lfloor x \rfloor) + \lfloor x \rfloor,$$

where $f^{-1}(x - \lfloor x \rfloor)$ is the representative in the interval $[0, 1)$. We note that

$$\lfloor f^{-1}(x - \lfloor x \rfloor) + \lfloor x \rfloor \rfloor = \lfloor x \rfloor$$

and

$$\lfloor f(x - \lfloor x \rfloor) + \lfloor x \rfloor \rfloor = \lfloor x \rfloor.$$

Fig. 4.2 An
orientation-preserving
homeomorphism

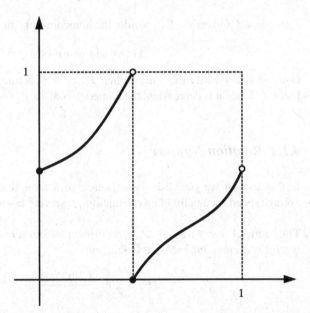

Thus,

$$(F \circ H)(x) = F\left(f^{-1}(x - \lfloor x \rfloor) + \lfloor x \rfloor\right)$$
$$= f\left(f^{-1}(x - \lfloor x \rfloor)\right) + \lfloor x \rfloor$$
$$= x - \lfloor x \rfloor + \lfloor x \rfloor = x$$

and

$$(H \circ F)(x) = H\left(f(x - \lfloor x \rfloor) + \lfloor x \rfloor\right)$$
$$= f^{-1}\left(f(x - \lfloor x \rfloor)\right) + \lfloor x \rfloor$$
$$= x - \lfloor x \rfloor + \lfloor x \rfloor = x$$

for $x \in \mathbb{R}$. This shows that H is the inverse of F. Hence, F is a homeomorphism. $\square$

Now we consider the class of orientation-preserving homeomorphisms.

Definition 4.2 A homeomorphism $f : S^1 \to S^1$ is said to be *orientation-preserving* if it has a lift which is an increasing function (see Fig. 4.2).

It follows from Proposition 4.1 that f is orientation-preserving if and only if all its lifts are increasing functions.

For example, the homeomorphisms of the circle considered in Examples 4.1 and 4.2 are orientation-preserving since the lifts in (4.3) and (4.5) are increasing functions. We also give an example of a homeomorphism that is not orientation-preserving.

Example 4.3 Given $\alpha \in \mathbb{R}$, consider the homeomorphism $f: S^1 \to S^1$ defined by

$$f(x) = -x + \alpha \bmod 1.$$

One can easily verify that the function $F: \mathbb{R} \to \mathbb{R}$ defined by $F(x) = -x + \alpha$ is a lift of f. Since it is decreasing, the homeomorphism f is not orientation-preserving.

4.1.2 Rotation Number

In this section we establish the existence of a limit that can be described as an average speed for any lift of an orientation-preserving homeomorphism of the circle.

Theorem 4.1 *Let $f: S^1 \to S^1$ be an orientation-preserving homeomorphism. If F is a lift of f, then for each $x \in \mathbb{R}$ the limit*

$$\rho(F) = \lim_{n \to \infty} \frac{F^n(x) - x}{n} \in \mathbb{R}_0^+ \tag{4.9}$$

exists and is independent of x. Moreover, if G is another lift of f, then

$$\rho(G) - \rho(F) \in \mathbb{Z}.$$

Proof We first assume that $F(x) > x$ for every $x \in \mathbb{R}$. Given a point $x \in \mathbb{R}$, consider the sequence $a_n = F^n(x) - x$. For each $m, n \in \mathbb{N}$, we have

$$a_{m+n} = F^{m+n}(x) - x$$
$$= F^m(F^n(x)) - F^n(x) + a_n. \tag{4.10}$$

Since

$$\lfloor a_n \rfloor \le F^n(x) - x < \lfloor a_n \rfloor + 1, \tag{4.11}$$

we obtain

$$F^m(F^n(x)) < F^m(x + \lfloor a_n \rfloor) + 1. \tag{4.12}$$

On the other hand, we have

$$F^m(x + \lfloor a_n \rfloor) - (x + \lfloor a_n \rfloor) = F^m(x) - x = a_m$$

and it follows from (4.10) and (4.12) that

$$a_{m+n} < F^m(x + \lfloor a_n \rfloor) + 1 - F^n(x) + a_n$$
$$= a_m + a_n + x + \lfloor a_n \rfloor - F^n(x) + 1.$$

Finally, by (4.11), we obtain

$$a_{m+n} \le a_m + a_n + 1$$

and the sequence $c_n = a_n + 1$ satisfies condition (3.25) in Lemma 3.2. Hence, the limit

$$\lim_{n \to \infty} \frac{F^n(x) - x}{n} = \lim_{n \to \infty} \frac{a_n}{n} = \inf\left\{\frac{a_n}{n} : n \in \mathbb{N}\right\} \qquad (4.13)$$

exists. Since

$$a_n = F^n(x) - x > 0$$

(recall that F is increasing), the limit in (4.13) is finite.

Now we show that the limit in (4.13) is independent of x. Given $x, y \in \mathbb{R}$ and $k \in \mathbb{N}$ with $|x - y| \le k$, we have

$$F(x) \le F(y + k) = F(y) + k$$

and

$$F(x) \ge F(y - k) = F(y) - k.$$

Hence, $|F(x) - F(y)| \le k$ and it follows by induction that

$$\left|F^n(x) - F^n(y)\right| \le k \quad \text{for } n \in \mathbb{N}.$$

This implies that

$$\left|\frac{F^n(x) - x}{n} - \frac{F^n(y) - y}{n}\right| = \left|\frac{F^n(x) - F^n(y)}{n} + \frac{y - x}{n}\right| \le \frac{2k}{n} \to 0$$

when $n \to \infty$ and thus,

$$\lim_{n \to \infty} \frac{F^n(x) - x}{n} = \lim_{n \to \infty} \frac{F^n(y) - y}{n}$$

for $x, y \in \mathbb{R}$ (given $x, y \in \mathbb{R}$, one can always choose $k \in \mathbb{N}$ such that $|x - y| \le k$).

It remains to establish the last property in the theorem. By Proposition 4.1, if F and G are lifts of f, then there exists a $k \in \mathbb{Z}$ such that $G - F = k$. It follows by induction that

$$G^n(x) = F^n(x) + nk.$$

Therefore,

$$\rho(G) = \lim_{n \to \infty} \frac{G^n(x) - x}{n}$$

$$= \lim_{n \to \infty} \frac{F^n(x) - x}{n} + k = \rho(F) + k,$$

which establishes the last property in the theorem.   □

For each $x \in \mathbb{R}$, we also have

$$\rho(F) = \lim_{n \to \infty} \frac{F^n(x)}{n}.$$

Now we introduce the notion of the rotation number.

Definition 4.3 The *rotation number* of an orientation-preserving homeomorphism $f: S^1 \to S^1$ is defined by

$$\rho(f) = \pi(\rho(F)), \tag{4.14}$$

where F is any lift of f and where $\pi(x) = [x]$.

It follows from the last property in Theorem 4.1 that the rotation number is well defined, that is, $\rho(f)$ does not depend on the lift F used in (4.14).

Example 4.4 Given $\alpha \in \mathbb{R}$, consider the rotation R_α in (4.2). For the lift F in (4.3), we obtain

$$\frac{F^n(x) - x}{n} = \frac{x + n(\alpha + k) - x}{n} = \alpha + k$$

and thus, $\rho(F) = \alpha + k$. Hence,

$$\rho(R_\alpha) = \pi(\rho(F)) = \alpha \bmod 1.$$

Example 4.5 Now we consider the homeomorphism $f: S^1 \to S^1$ defined in (4.4), with $|\beta| < 1/(2\pi)$. Since the limit in (4.9) does not depend on x, for the lift F in (4.5), we obtain

$$\rho(F) = \lim_{n \to \infty} \frac{F^n(0) - 0}{n} = 0.$$

4.1.3 Rational Rotation Number

Here and in the next section, we verify that the properties of an orientation-preserving homeomorphism of the circle strongly depend on whether the rotation number is rational or irrational.

In this section we consider the homeomorphisms with rational rotation number. We recall that $x \in S^1$ is said to be a *periodic point* of a map $f: S^1 \to S^1$ if $f^q(x) = x$ for some $q \in \mathbb{N}$.

Theorem 4.2 *Let $f: S^1 \to S^1$ be an orientation-preserving homeomorphism. Then $\rho(f) \in \mathbb{Q}$ if and only if f has at least one periodic point.*

Proof We first assume that $\rho(f) = 0$ and we show that f has a fixed point. Otherwise, if f has no fixed points and F is a lift of f, then

$$F(x) - x \in \mathbb{R} \setminus \mathbb{Z} \qquad (4.15)$$

for $x \in \mathbb{R}$. Indeed, if $F(x) - x \in \mathbb{Z}$ for some $x \in \mathbb{R}$, then

$$\pi(x) = \pi(F(x)) = f(\pi(x))$$

and thus, $\pi(x)$ would be a fixed point of f. Since F is continuous, it follows from (4.15) that there exists a $k \in \mathbb{Z}$ such that

$$k < F(x) - x < k + 1 \quad \text{for } x \in \mathbb{R}. \qquad (4.16)$$

On the other hand,

$$F(x + 1) - (x + 1) = F(x) - x \qquad (4.17)$$

for $x \in \mathbb{R}$ and thus, the continuous function $x \mapsto F(x) - x$ is completely determined by its values on the compact interval $[0, 1]$. Hence, it follows from (4.16) and Weierstrass' theorem that there exists an $\varepsilon > 0$ such that

$$k + \varepsilon \le F(x) - x \le k + 1 - \varepsilon \qquad (4.18)$$

for $x \in \mathbb{R}$. Since

$$F^n(x) - x = \sum_{i=0}^{n-1} \left[F(F^i(x)) - F^i(x) \right],$$

it follows from (4.18) that

$$k + \varepsilon \le \frac{F^n(x) - x}{n} \le k + 1 - \varepsilon$$

and thus,

$$\rho(f) = \lim_{n \to \infty} \frac{F^n(x) - x}{n} \bmod 1 \in [\varepsilon, 1 - \varepsilon].$$

This contradicts the hypothesis that $\rho(f) = 0$ and thus, f must have a fixed point.

Now we assume that $\rho(f) = p/q \in \mathbb{Q}$. Since F^q is a lift of f^q, we obtain

$$\rho(f^q) = \lim_{n \to \infty} \frac{(F^q)^n(x) - x}{n} \bmod 1$$

$$= q \lim_{n \to \infty} \frac{F^{qn}(x) - x}{qn} \bmod 1$$

$$= q\rho(f) \bmod 1$$

$$= p \bmod 1 = 0.$$

It follows from the above argument for a zero rotation number that the homeomorphism f^q has a fixed point, which is a periodic point of f.

For the converse statement, we assume that f has a periodic point. Then there exist $y \in \mathbb{R}$ and $q \in \mathbb{N}$ such that

$$f^q\big(\pi(y)\big) = \pi(y).$$

It follows from (4.1) by induction that $f^q \circ \pi = \pi \circ F^q$ and thus,

$$\pi\big(F^q(y)\big) = \pi(y).$$

Hence, $F^q(y) = y + p$ for some $p \in \mathbb{Z}$. On the other hand, it follows from (4.17) that

$$F(x + p) = F(x) + p$$

for $x \in \mathbb{R}$ and thus, we also have

$$F^q(x + p) = F^q(x) + p \qquad (4.19)$$

for $x \in \mathbb{R}$ and $q \in \mathbb{N}$. In particular, taking $x = y$, we obtain

$$
\begin{aligned}
F^{2q}(y) = F^q\big(F^q(y)\big) &= F^q(y + p) \\
&= F^q(y) + p = y + 2p
\end{aligned}
$$

and it follows by induction that

$$F^{nq}(y) = y + np \quad \text{for } n \in \mathbb{N}.$$

Thus,

$$
\begin{aligned}
\rho(F) &= \lim_{n \to \infty} \frac{F^{nq}(y) - y}{nq} \\
&= \lim_{n \to \infty} \frac{np}{nq} = \frac{p}{q}.
\end{aligned}
$$

This completes the proof of the theorem. $\square$

We continue to consider a homeomorphism $f \colon S^1 \to S^1$. We recall that, given $q \in \mathbb{N}$, a point $x \in S^1$ is said to be a q-periodic point of f if $f^q(x) = x$. It follows from the proof of Theorem 4.2 that f^q has a fixed point, that is, f has a q-periodic point if and only if $\rho(f) = p/q$ for some $p \in \mathbb{N}$. Thus, f has a periodic point with period q if and only if $\rho(f) = p/q$ with p and q coprime. Indeed, by the previous observation, f has no l-periodic points for any $l < q$.

We also have the following result.

Theorem 4.3 *Let $f \colon S^1 \to S^1$ be an orientation-preserving homeomorphism. If $\rho(f) = p/q$ with p and q coprime, then all periodic points of f have period q.*

Proof Let $x \in S^1$ be a periodic point of f. It follows from the former discussion that x has period $l = dq$ for some $d \in \mathbb{N}$. On the other hand, it follows from the proof of Theorem 4.2 that if F is a lift of f, then

$$F^l(x) = x + dp + ml \qquad (4.20)$$

for some $m \in \mathbb{Z}$. In fact, one can always assume that $m = 0$. Indeed, if G is another lift of f, then $F = G + m$ for some $m \in \mathbb{Z}$ and thus, $F^l = G^l + ml$. Hence, it is sufficient to replace F by G.

Now we show that $F^q(x) = x + p$. Since F is increasing, if $F^q(x) > x + p$, then it follows from (4.19) that

$$F^{2q}(x) > F^q(x + p) = F^q(x) + p > x + 2p$$

and by induction,

$$F^l(x) = F^{dq}(x) > x + dp.$$

This contradicts (4.20) (with $m = 0$). We obtain in an analogous manner a contradiction when $F^q(x) < x + p$. Thus, $F^q(x) = x + p$ and the point x has period q. $\square$

4.1.4 Irrational Rotation Number

In this section we consider the homeomorphisms of the circle with irrational rotation number. We first show that the orbits of these homeomorphisms are ordered as the orbits of the rotation R_ρ, where ρ is the rotation number.

Theorem 4.4 *Let F be a lift of an orientation-preserving homeomorphism of the circle $f \colon S^1 \to S^1$ with $\rho(f) \in \mathbb{R} \setminus \mathbb{Q}$. For each $x \in \mathbb{R}$ and $n_1, n_2, m_1, m_2 \in \mathbb{Z}$, we have*

$$F^{n_1}(x) + m_1 < F^{n_2}(x) + m_2 \qquad (4.21)$$

if and only if

$$n_1 \rho(F) + m_1 < n_2 \rho(F) + m_2. \qquad (4.22)$$

Proof It is sufficient to take $n_1 \neq n_2$ since otherwise there is nothing to prove.

We first assume that (4.21) holds. For $n_1 > n_2$, we have

$$F^{n_1 - n_2}(x) < x + m_2 - m_1$$

for $x \in \mathbb{R}$. Thus,

$$F^{2(n_1 - n_2)}(x) < F^{n_1 - n_2}(x) + m_2 - m_1 < x + 2(m_2 - m_1)$$

and by induction,

$$F^{n(n_1-n_2)}(x) < x + n(m_1 - m_2).$$

We obtain

$$\rho(F) = \lim_{n\to\infty} \frac{F^{n(n_1-n_2)}(x) - x}{n(n_1 - n_2)} < \frac{m_2 - m_1}{n_1 - n_2},$$

with strict inequality since $\rho(f)$ is irrational. This shows that inequality (4.22) holds. Analogously, for $n_1 < n_2$, we have

$$F^{n_2-n_1}(x) > x + m_1 - m_2$$

for $x \in \mathbb{R}$ and thus,

$$F^{n(n_2-n_1)}(x) > x + n(m_1 - m_2).$$

Hence,

$$\rho(F) = \lim_{n\to\infty} \frac{F^{n(n_2-n_1)}(x) - x}{n(n_2 - n_1)} > \frac{m_1 - m_2}{n_2 - n_1}$$

and inequality (4.22) also holds in this case.

In the other direction, we must show that if

$$F^{n_1}(x) + m_1 \geq F^{n_2}(x) + m_2,$$

then

$$n_1\rho(F) + m_1 \geq n_2\rho(F) + m_2.$$

But since $\rho(f)$ is irrational, none of these inequalities can be an equality. Thus, this is equivalent to show that if

$$F^{n_1}(x) + m_1 > F^{n_2}(x) + m_2,$$

then

$$n_1\rho(F) + m_1 > n_2\rho(F) + m_2.$$

For this it is sufficient to reverse all inequalities in the previous argument. $\qquad\square$

Now we establish a more precise relation between a homeomorphism of the circle with irrational rotation number ρ and the rotation of the circle R_ρ.

Theorem 4.5 *Let $f\colon S^1 \to S^1$ be an orientation-preserving homeomorphism with rotation number $\rho(f) \in \mathbb{R} \setminus \mathbb{Q}$. Then there exists a nondecreasing and onto continuous function $h\colon S^1 \to S^1$ such that $h \circ f = R_{\rho(f)} \circ h$.*

Proof Given a lift F of the homeomorphism f and a point $x \in \mathbb{R}$, consider the sets

$$A = \{F^n(x) + m : n, m \in \mathbb{Z}\} \quad \text{and} \quad B = \{n\rho + m : n, m \in \mathbb{Z}\}, \tag{4.23}$$

where $\rho = \rho(F)$. We define a function $H : \mathbb{R} \to \mathbb{R}$ by

$$H(y) = \sup\{n\rho + m : F^n(x) + m \le y\}. \tag{4.24}$$

It follows from Theorem 4.4 that H is nondecreasing. Moreover, H is constant on each interval contained in the complement of $\overline{A}$. Indeed, if $[a, b] \subset S^1 \setminus \overline{A}$, then

$$F^n(x) + m \le a \quad \Leftrightarrow \quad F^n(x) + m \le b$$

for every $n, m \in \mathbb{Z}$ and thus $H(a) = H(b)$.

Lemma 4.1 *The set B is dense in $\mathbb{R}$.*

Proof Since $y \in B$ if and only if $y + m \in B$ for some $m \in \mathbb{Z}$, it suffices to show that $B \cap [0, 1]$ is dense in $[0, 1]$. Clearly, the set $B \cap [0, 1]$ is infinite. Otherwise, there would exist pairs $(n_1, m_1) \ne (n_2, m_2)$ in $\mathbb{Z}^2$ such that

$$n_1\rho + m_1 = n_2\rho + m_2,$$

but this is impossible since ρ is irrational (if $n_1 = n_2$, then $m_1 \ne m_2$). Let then x_n be a sequence in $B \cap [0, 1]$ with infinitely many values. Since $[0, 1]$ is compact, one can assume that the sequence x_n is convergent. Hence, given $\varepsilon > 0$, there exist $m, n \in \mathbb{N}$ such that $0 < |x_n - x_m| < \varepsilon$. Writing

$$x_n = n_1\rho + m_1 \quad \text{and} \quad x_m = n_2\rho + m_2,$$

we obtain

$$x_n - x_m = (n_1 - n_2)\rho + (m_1 - m_2) \in B.$$

This shows that the set $B \supset \{k(x_n - x_m) : k \in \mathbb{Z}\}$ is ε-dense in $\mathbb{R}$. Since ε is arbitrary, we conclude that B is dense in $\mathbb{R}$. $\square$

Since ρ is irrational, it follows from Theorem 4.4 that

$$H\big(F^n(x) + m\big) = n\rho + m. \tag{4.25}$$

This implies that the function H has no jumps. Indeed, by (4.25), we have

$$H(\mathbb{R}) \supset H(A) = B$$

and by Lemma 4.1, the set B is dense in $\mathbb{R}$. Since H is monotonous, this implies that it is also continuous.

Now we consider the lift $S : \mathbb{R} \to \mathbb{R}$ of R_ρ given by $S(x) = x + \rho$. By (4.25), we have

$$(H \circ F)\big(F^n(x) + m\big) = H\big(F^{n+1}(x) + m\big) = (n + 1)\rho + m$$

and

$$(S \circ H)\big(F^n(x) + m\big) = S(n\rho + m) = (n+1)\rho + m.$$

Thus,

$$H \circ F = S \circ H \quad \text{in } A. \tag{4.26}$$

Since the maps H, F and S are continuous, identity (4.26) holds in $\overline{A}$ and thus also in $\mathbb{R}$ (recall that H is constant on each interval contained in the complement of $\overline{A}$). That is, we have

$$H \circ F = S \circ H \quad \text{in } \mathbb{R}. \tag{4.27}$$

On the other hand,

$$
\begin{aligned}
H(y+1) &= \sup\big\{n\rho + m : F^n(x) + m \le y + 1\big\} \\
&= \sup\big\{n\rho + m : F^n(x) + m - 1 \le y\big\} \\
&= \sup\big\{n\rho + m - 1 : F^n(x) + m - 1 \le y\big\} + 1 \\
&= H(y) + 1.
\end{aligned}
$$

The function H is also onto. Indeed, since it is continuous, we have

$$H(\mathbb{R}) = H\big([0, 1]\big) \supset \overline{B} = \mathbb{R}.$$

Hence, the function $h \colon S^1 \to S^1$ defined by

$$h(y) = H(y) \bmod 1$$

is continuous, nondecreasing and onto. Moreover, it follows from property (4.27) that $h \circ f = R_\rho \circ h$. $\qquad\square$

If the homeomorphism has a dense positive semiorbit, which by Theorem 3.2 is equivalent to the existence of a dense orbit, then Theorem 4.5 can be strengthened as follows.

Theorem 4.6 (Poincaré) *Let $f \colon S^1 \to S^1$ be an orientation-preserving homeomorphism with $\rho(f) \in \mathbb{R} \setminus \mathbb{Q}$. If f has a dense positive semiorbit, then it is topologically conjugate to the rotation $R_{\rho(f)}$, that is, there exists a homeomorphism $h \colon S^1 \to S^1$ such that $h \circ f = R_{\rho(f)} \circ h$.*

Proof Let $x \in S^1$ be a point whose positive semiorbit is dense in S^1. Now consider the function $h \colon S^1 \to S^1$ constructed in Theorem 4.5 taking the point x in (4.23) and (4.24). Now the set A is dense in S^1 and thus, the function H in (4.24) is bijective (we recall that H is constant on each interval contained in $\mathbb{R} \setminus \overline{A}$, which now is the empty set). Thus, the function h is also bijective. It remains to show that h is open, that is, that the image $h(U)$ of an open set U is also open. Since h is continuous, it maps compact sets to compact sets. Hence, given an open set U, the image $h(S^1 \setminus U) = S^1 \setminus h(U)$ is compact and thus, $h(U)$ is an open set. This shows that h is a homeomorphism.   □

4.2 Diffeomorphisms of the Circle

In this section we consider the particular case of the diffeomorphisms of the circle (we recall that a diffeomorphism is a bijective differentiable map with differentiable inverse). We show that any sufficiently regular diffeomorphism $f\colon S^1 \to S^1$ with irrational rotation number is topologically conjugate to a rotation. More precisely, there exists a homeomorphism $h\colon S^1 \to S^1$ such that

$$h \circ f = R_{\rho(f)} \circ h.$$

We first recall that a function $\varphi\colon S^1 \to \mathbb{R}$ is said to have *bounded variation* if

$$\mathrm{Var}(\varphi) = \sup \sum_{k=1}^{n} \left| \varphi(x_k) - \varphi(y_k) \right| < +\infty,$$

where the supremum is taken over all disjoint open intervals $(x_1, y_1), \ldots, (x_n, y_n)$, with $n \in \mathbb{N}$.

Example 4.6 Let $\varphi\colon S^1 \to \mathbb{R}$ be a differentiable function with bounded derivative. Then there exists a $K > 0$ such that $|\varphi'(x)| \le K$ for $x \in S^1$. If (x_i, y_i), for $i = 1, \ldots, n$, are disjoint open intervals with $y_1 \le x_2$, $y_2 \le x_3, \ldots, y_{n-1} \le x_n$, then

$$\sum_{i=1}^{n} \left| \varphi(y_i) - \varphi(x_i) \right| = \sum_{i=1}^{n} \left| \varphi'(z_i) \right| (y_i - x_i)$$

$$\le \sum_{i=1}^{n} K(y_i - x_i) \le K,$$

where z_i is some point in the interval (x_i, y_i). Thus, $\mathrm{Var}(\varphi) \le K$ and φ has bounded variation.

The following result gives conditions for a diffeomorphism of the circle to be topologically conjugate to a rotation.

Theorem 4.7 (Denjoy) *Let* $f\colon S^1 \to S^1$ *be an orientation-preserving* C^1 *diffeomorphism whose derivative has bounded variation. If* $\rho(f) \in \mathbb{R} \setminus \mathbb{Q}$, *then* f *is topologically conjugate to the rotation* $R_{\rho(f)}$.

Proof By Theorem 4.6, it suffices to show that there exists a point $z \in S^1$ whose positive semiorbit is dense, which is equivalent to $\omega(z) = S^1$. If $\omega(z) \ne S^1$, then the set $S^1 \setminus \omega(z)$ is a disjoint union of maximal intervals (we say that an open interval $I \subset S^1 \setminus \omega(z)$ is *maximal* if any nonempty open interval J such that $I \subset J \subset S^1 \setminus \omega(z)$ coincides with I). Moreover, since f is a homeomorphism, the set $\omega(z)$ is f-invariant and thus, the image and the preimage of any of these intervals are also maximal intervals.

Now let $I \subset S^1 \setminus \omega(z)$ be a maximal interval. We show that the sets $f^n(I)$, for $n \in \mathbb{Z}$, are pairwise disjoint. By the former paragraph, if there exist integers $m > n$ such that $f^m(I) \cap f^n(I) \neq \varnothing$, then

$$f^{m-n}(I) \cap I \neq \varnothing$$

and thus $f^{m-n}(I) = I$. Since f is continuous, we also have $f^{m-n}(\overline{I}) = \overline{I}$.

Lemma 4.2 *Let $g: J \to J$ be a continuous function on some interval $J \subset \mathbb{R}$. If $K \subset J$ is a compact interval such that $g(K) \supset K$, then g has a fixed point in K.*

Proof Write $K = [\alpha, \beta]$. Since $g(K) \supset K$, there exist $a, b \in K$ such that

$$g(a) = \alpha \leq a \quad \text{and} \quad g(b) = \beta \geq b.$$

Since $g(a) - a \leq 0$ and $g(b) - b \geq 0$, the continuous function $x \mapsto g(x) - x$ has a zero in K.   □

It follows from the lemma that f^{m-n} has a fixed point in $\overline{I}$, but this is impossible since the rotation number is irrational. Thus, the intervals $f^n(I)$ are pairwise disjoint and their lengths λ_n satisfy

$$\sum_{n\in\mathbb{Z}} \lambda_n \leq 1. \tag{4.28}$$

Now we establish some auxiliary results.

Lemma 4.3 *There exist infinitely many $n \in \mathbb{N}$ such that for each $x \in S^1$ the intervals $J = (x, f^{-n}(x))$, $f(J), \ldots, f^n(J)$ are pairwise disjoint.*

Proof For each $k = 0, \ldots, n$, we have

$$f^k(J) = \left(f^k(x), f^{k-n}(x) \right)$$

since f is orientation-preserving. Hence, the intervals $f^k(J)$ are pairwise disjoint if and only if $f^k(x), f^{k-n}(x) \notin f^l(J)$ for $k, l = 0, \ldots, n$ with $l < k$, or equivalently,

$$f^k(x) \notin J \quad \text{for } |k| \leq n.$$

We note that this property only depends on the ordering of the orbit of x. By Theorem 4.4, this is the same as the ordering of the orbits of the rotation R_ρ, where

$\rho = \rho(f)$. Since ρ is irrational, all negative semiorbits are dense. Thus, there exist infinitely many $n \in \mathbb{N}$ such that

$$R_\rho^k(y) \notin \left(y, R_\rho^{-n}(y)\right) \quad \text{for } |k| \le n \text{ and } y \in S^1.$$

This yields the desired result. □

Lemma 4.4 *If $J \subset S^1$ is an open interval such that the sets $J, f(J), \ldots, f^{n-1}(J)$ are pairwise disjoint, then*

$$c^{-1} \le \frac{(f^n)'(y)}{(f^n)'(z)} \le c \tag{4.29}$$

for any $y, z \in \overline{J}$, where $c = \exp \operatorname{Var}(\log f') < +\infty$.

Proof We define a function $\varphi \colon S^1 \to \mathbb{R}$ by $\varphi = \log f'$ (f is orientation-preserving and hence $f' > 0$). Since the sets $J, \ldots, f^{n-1}(J)$ are pairwise disjoint, given $y, z \in \overline{J}$, the open intervals determined by the pairs of points $f^k(y)$ and $f^k(z)$, for $k = 0, \ldots, n-1$, are also disjoint. Thus,

$$\operatorname{Var}(\varphi) \ge \sum_{k=0}^{n-1} \left| \varphi(f^k(y)) - \varphi(f^k(z)) \right|$$

$$\ge \left| \sum_{k=0}^{n-1} \varphi(f^k(y)) - \varphi(f^k(z)) \right|$$

$$= \left| \log \prod_{k=0}^{n-1} f'(f^k(y)) - \log \prod_{k=0}^{n-1} f'(f^k(z)) \right|$$

$$= \left| \log \frac{(f^n)'(y)}{(f^n)'(z)} \right|.$$

This implies that

$$-\operatorname{Var}(\varphi) \le \log \frac{(f^n)'(y)}{(f^n)'(z)} \le \operatorname{Var}(\varphi),$$

which yields inequality (4.29) provided that $\operatorname{Var}(\varphi)$ is finite. Since S^1 is compact and f' is continuous, we have $\inf f' > 0$. Hence,

$$\left| \varphi(y) - \varphi(z) \right| = \left| \log f'(y) - \log f'(z) \right| \le \frac{|f'(y) - f'(z)|}{\inf f'}$$

for $x, y \in S^1$ and since f' has bounded variation, we obtain

$$\operatorname{Var}(\varphi) \le \frac{\operatorname{Var}(f')}{\inf f'} < +\infty.$$

This completes the proof of the lemma. □

Applying Lemma 4.4 to the intervals $J = (x, f^{-n}(x))$ in Lemma 4.3, with $y = x \in I$ and $z = f^{-n}(x)$ (with n independent of x), we conclude that

$$c^{-1} \le (f^n)'(x)(f^{-n})'(x) \le c.$$

Since

$$a + b \ge \sqrt{ab} \quad \text{for } a, b \ge 0,$$

we obtain

$$
\begin{aligned}
\lambda_n + \lambda_{-n} &= \int_I (f^n)'(x)\, dx + \int_I (f^{-n})'(x)\, dx \\
&= \int_I \left[(f^n)'(x) + (f^{-n})'(x) \right] dx \\
&\ge \int_I \sqrt{(f^n)'(x)(f^{-n})'(x)}\, dx \\
&\ge c^{-1/2} \lambda_0,
\end{aligned}
$$

for the integers n given by Lemma 4.3. This implies that

$$\sum_{m \in \mathbb{Z}} \lambda_m = +\infty,$$

which contradicts (4.28). Thus, there exists a point $z \in S^1$ with $\omega(z) = S^1$.   □

4.3 Maps of the Interval

In this section we consider the class of continuous maps of a compact interval. In particular, we study the properties of their periodic points. We also establish Sharkovsky's theorem, which describes how the existence of periodic points with a given period determines the existence of periodic points with another period.

4.3.1 Existence of Periodic Points

Let $f\colon I \to I$ be a continuous map of an interval $I \subset \mathbb{R}$.

Definition 4.4 Given intervals $J, K \subset I$ such that $f(J) \supset K$, we say that J *covers* K and we write $J \to K$.

This notion can be used in the study of the existence of periodic points.

Proposition 4.2 *Let* $f : I \to I$ *be a continuous map of a compact interval* $I \subset \mathbb{R}$. *If there exist closed intervals* $I_0, I_1, \ldots, I_{n-1} \subset I$ *such that*

$$I_0 \to I_1 \to I_2 \to \cdots \to I_{n-1} \to I_0,$$

then f *has an* n-*periodic point* $x \in I$ *such that* $f^m(x) \in I_m$ *for* $m = 0, 1, \ldots, n-1$.

Proof We first show that there exists a closed interval $J_0 \subset I_0$ such that $f(J_0) = I_1$. Since $f(I_0) \supset I_1$, there exist points $a_0, b_0 \in I_0$ whose images are the endpoints of I_1. If J_0 is the closed interval with endpoints a_0 and b_0, then $f(J_0) = I_1$.

Now let us assume that we constructed closed intervals

$$J_0 \supset J_1 \supset \cdots \supset J_{m-1}$$

contained in I_0, for some $m < n$, such that $f^{k+1}(J_k) = I_{k+1}$ for $k = 0, \ldots, m-1$. Then

$$f^{m+1}(J_{m-1}) = f(I_m) \supset I_{m+1}$$

and an analogous argument shows that there exists a closed interval $J_m \subset J_{m-1}$ such that $f^{m+1}(J_m) = I_{m+1}$. Thus, we obtain closed intervals

$$J_0 \supset J_1 \supset \cdots \supset J_{n-1}$$

such that $f^{k+1}(J_k) = I_{k+1}$ for $k = 0, \ldots, n-1$, where $I_n = I_0$. In particular,

$$f^n(J_{n-1}) = I_0 \supset J_{n-1} \tag{4.30}$$

and each point $x \in J_{n-1}$ satisfies

$$f^m(x) \in f^m(J_{n-1}) \subset f^m(J_{m-1}) = I_m \tag{4.31}$$

for $m = 0, \ldots, n-1$. On the other hand, it follows from (4.30) and Lemma 4.2 that f^n has a fixed point in J_{n-1}. Thus, f has an n-periodic point in J_{n-1}, which also satisfies (4.31). $\qquad\square$

Now we consider a quadratic map.

Example 4.7 Given $a > 4$, consider the map $f : [0, 1] \to \mathbb{R}$ defined by

$$f(x) = ax(1 - x).$$

We have

$$f\left(\left[\frac{1}{a}, \frac{1}{2}\right]\right) = \left[1 - \frac{1}{a}, \frac{a}{4}\right] \supset \left[1 - \frac{1}{a}, 1\right]$$

and

$$f\left(\left[1 - \frac{1}{a}, 1\right]\right) = \left[0, 1 - \frac{1}{a}\right] \supset \left[\frac{1}{a}, \frac{1}{2}\right].$$

Since

$$\left[\frac{1}{a}, \frac{1}{2}\right] \cap \left[1 - \frac{1}{a}, 1\right] = \varnothing,$$

it follows from Proposition 4.2 that f has a periodic point in $[1/a, 1/2]$ with period 2.

The criterion in Proposition 4.2 can be used to establish the following particular case of Sharkovsky's theorem (Theorem 4.9).

Theorem 4.8 *Let $f : I \to I$ be a continuous map of a compact interval $I \subset \mathbb{R}$. If f has a periodic point with period 3, then it has periodic points with all periods.*

Proof Let $x_1 < x_2 < x_3$ be the elements of the orbit of a periodic point with period 3. We first assume that $f(x_2) = x_3$. We have $f^2(x_2) = x_1$ and thus,

$$[x_1, x_2] \leftrightarrow [x_2, x_3] \circlearrowleft .$$

On the other hand, if $f(x_2) = x_1$, then

$$[x_2, x_3] \leftrightarrow [x_1, x_2] \circlearrowleft .$$

In both cases, we have $I \to I$ taking, respectively, $I = [x_2, x_3]$ or $I = [x_1, x_2]$. It follows from Proposition 4.2 that f has a fixed point.

Furthermore, given an integer $n \geq 2$ with $n \neq 3$, we have

$$I_1 \to I_2 \to I_2 \to \cdots \to I_2 \to I_2 \to I_1, \qquad (4.32)$$

with $n + 1$ elements, taking, respectively,

$$I_1 = [x_1, x_2] \quad \text{and} \quad I_2 = [x_2, x_3]$$

or

$$I_1 = [x_2, x_3] \quad \text{and} \quad I_2 = [x_1, x_2].$$

It follows from Proposition 4.2 that f has an n-periodic point $x \in I_1$. If it did not have period n, then $x \in I_1 \cap I_2 = \{x_2\}$, that is, $x = x_2$. The orbit of x_2 belongs successively to the intervals

$$I_1 \, I_2 \, I_2 \, I_1 \, I_2 \, I_2 \, I_1 \cdots$$

and thus, it cannot belong successively to the intervals in (4.32) unless $n = 3$ (but we took $n \neq 3$). This contradiction shows that the periodic point x has period n. $\square$

Fig. 4.3 A periodic point
with period 3

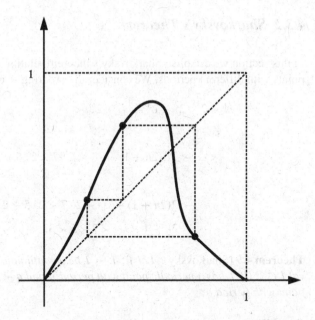

Fig. 4.4 A periodic point
with period 5

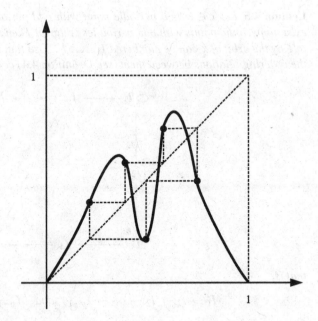

Figures 4.3 and 4.4 give examples on the interval [0, 1] with periods 3 and 5, respectively.

4.3.2 Sharkovsky's Theorem

In this section we establish Sharkovsky's theorem relating the existence of periodic points with different periods. We consider the ordering $\prec$ on $\mathbb{N}$ defined by

$$1 \prec 2 \prec 2^2 \prec 2^3 \prec \cdots \prec 2^m \prec \cdots$$

$$\cdots$$

$$\prec \cdots \prec 2^m(2n+1) \prec \cdots \prec 2^m 7 \prec 2^m 5 \prec 2^m 3 \prec \cdots$$

$$\cdots$$

$$\prec \cdots \prec 2(2n+1) \prec \cdots \prec 2 \cdot 7 \prec 2 \cdot 5 \prec 2 \cdot 3 \prec \cdots$$

$$\prec \cdots \prec 2n+1 \prec \cdots \prec 7 \prec 5 \prec 3.$$

Theorem 4.9 (Sharkovsky) *Let $f \colon I \to I$ be a continuous map of a compact interval $I \subset \mathbb{R}$. If f has a periodic point with period p and $q \prec p$, then f has a periodic point with period q.*

Proof We first establish two auxiliary results.

Lemma 4.5 *Let $x \in I$ be a periodic point with odd period $p > 1$ such that there exist no periodic points with odd period less than p. Then the intervals determined in I by the orbit of x can be numbered $I_1, \ldots, I_{p-1}$ so that the graph obtained from the covering relations between them (see Definition 4.4) contains the subgraph*

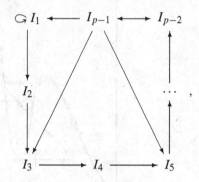

that is,

$$I_1 \to I_1 \to I_2 \to \cdots \to I_{p-1} \quad \text{and} \quad I_{p-1} \to I_k$$

for any odd k.

Proof Consider the interval $I_1 = [u, v]$, where

$$u = \max\{y \in \gamma(x) : f(y) > y\} \quad \text{and} \quad v = \min\{y \in \gamma(x) : y > u\}$$

(recall that $\gamma(x)$ is the orbit of x). By the definition of u, we have $f(v) < v$ (note that x is not a fixed point and thus $f(v) \neq v$). Moreover, $f(u) \geq v$ (since $f(u) > u$) and $f(v) \leq u$ (since $f(v) < v$). Therefore,

$$I_1 \to I_1. \tag{4.33}$$

Moreover, the inclusion $f(I_1) \supset I_1$ is proper (otherwise x would have period 2). Since

$$f^p(I_1) \supset f^{p-1}(I_1) \supset \cdots \supset f(I_1) \supset I_1$$

and x is p-periodic, we have $f^p(I_1) \supset \gamma(x)$ and so $f^p(I_1)$ contains all intervals determined by adjacent points in the orbit of x.

Now let $r = \operatorname{card} I^-$ and $s = \operatorname{card} I^+$, where

$$I^- = \gamma(x) \cap (-\infty, u] \quad \text{and} \quad I^+ = \gamma(x) \cap [v, +\infty).$$

Since $r + s = p$, we have $r \neq s$ (recall that p is odd). This implies that there exist adjacent points of $\gamma(x)$ in I^- or in I^+, thus determining an interval J, such that only one of them is mapped by f to the other interval. Otherwise, we would have $f(I^-) \subset I^+$ and $f(I^+) \subset I^-$ (since $f(u) > u$ and $f(v) < v$), but this is impossible since $r \neq s$. We also note that $J \to I_1$.

Now let

$$I_1 \to I_2 \to \cdots \to I_k \to I_1$$

be the shortest cycle of the form $I_1 \to \cdots \to I_1$ that is different from $I_1 \circlearrowleft$ (it follows from the former discussion that such a cycle always exists). Clearly, $k \leq p - 1$ since the orbit of x determines $p - 1$ intervals. Let q be the odd element of $\{k, k+1\}$. Since

$$I_1 \to \cdots \to I_k \to I_1 \quad \text{and} \quad I_1 \to \cdots \to I_k \to I_1 \to I_1,$$

it follows from Proposition 4.2 that f^q has a fixed point y. We note that y is not a fixed point of f. Otherwise,

$$y \in I_1 \cap \cdots \cap I_k \subset I_1 \cap I_2 \tag{4.34}$$

(recall that $k \geq 2$) would be in the orbit of x, which yields a contradiction since x is not a fixed point. It follows from the minimality of the odd period p that $q \geq p$ and thus $k = p - 1$. This shows that

$$I_1 \to I_2 \to \cdots \to I_{p-1} \to I_1 \tag{4.35}$$

is the shortest cycle of the form $I_1 \to \cdots \to I_1$ that is different from $I_1 \circlearrowleft$.

Now we show that $I_{p-1} \to I_k$ for k odd, which includes $I_{p-1} \to I_{p-2}$ since p is odd. We first verify that the intervals I_i are ordered in I in the form

$$I_{p-1}, I_{p-3}, \ldots, I_2, I_1, I_3, \ldots, I_{p-2} \tag{4.36}$$

(up to orientation). Since $I_1 \to \cdots \to I_{p-1} \to I_1$ is the shortest cycle of the form
$I_1 \to \cdots \to I_1$ that is different from $I_1 \circlearrowleft$, we conclude that if $I_k \to I_l$, then $l \le k + 1$ (or there would exist a shorter cycle of this form). This implies that I_1 only covers I_1 and I_2 (see (4.33) and (4.35)) and hence, I_2 is adjacent to I_1 (since $f(I_1)$ is connected). Since $I_1 = [u, v]$, we have $I_2 = [w, u]$, with $f(u) = v$ (recall that $f(u) > u$) and $f(v) = w$, or we have $I_2 = [v, w]$, with $f(u) = w$ and $f(v) = u$. We analyze only the first case since the second one is entirely analogous. Since $f(u) = v$ and I_2 does not cover I_1, we obtain $f(I_2) \subset [v, +\infty)$. But since I_2 covers I_3, we conclude that $I_3 = [v, t]$, with $t = f(w) = f^2(v)$ (since I_2 covers no other interval). Continuing this procedure yields the ordering in (4.36). This implies that

$$u_{p-1} < u_{p-3} < \cdots < u_2 < u < u_1 < u_3 < \cdots < u_{p-2},$$

where $u_i = f^i(u)$. Thus, we obtain $I_{p-1} = [u_{p-1}, u_{p-3}] \to I_k$ for k odd since $f(u_{p-1}) = u$ and $f(u_{p-3}) = u_{p-2}$. This completes the proof of the lemma. $\square$

Lemma 4.6 *If f has a periodic point with even period, then it has a periodic point with period 2.*

Proof Let x be a periodic point with even period $p > 2$. We consider two cases:

1. We first assume that there are no adjacent points in the orbit of x determining an interval $J \ne I_1$ that covers I_1. Let y and z be, respectively, the minimum and maximum of the orbit of x, that is,

$$y = \min \gamma(x) \quad \text{and} \quad z = \max \gamma(x).$$

   By construction, $f(u) \ge v$ and thus, $f([y, u])$ intersects $[v, +\infty)$. On the other hand, by hypothesis, the interval $[y, u]$ does not cover I_1 and thus, $f([y, u]) \subset [v, +\infty)$. One can show in an analogous manner that $f([v, z]) \subset (-\infty, u]$. Since f permutes the points in the orbit of x, we obtain

$$[y, u] \to [v, z] \to [y, u]$$

   and it follows from Proposition 4.2 that f has a periodic point with period 2.
2. Now we assume that there are adjacent points in the orbit of x determining an interval $I_k \ne I_1$ that covers I_1. If

$$I_1 \to \cdots \to I_k \to I_1$$

   is the shortest cycle of the form $I_1 \to \cdots \to I_1$ that is different from $I_1 \circlearrowleft$, then $k \le p - 1$. Now take $q \in \{k, k + 1\}$ even. Clearly $q \le p$. Since

$$I_1 \to \cdots \to I_k \to I_1 \quad \text{and} \quad I_1 \to \cdots \to I_k \to I_1 \to I_1,$$

   it follows from Proposition 4.2 that f^q has a fixed point y. We note that y is not a fixed point of f (see (4.34)). If p was the smallest even period, then $q = p$ and thus $k = p - 1$. Proceeding as in the proof of Lemma 4.5, one could then show that the intervals I_i must be ordered in I in the form

$$I_{p-2}, \ldots, I_2, I_1, I_3, \ldots, I_{p-1}$$

(up to orientation) and that $I_{p-1} \to I_k$ for k even. In particular, we would obtain the cycle $I_{p-1} \to I_{p-2} \to I_{p-1}$ and, by Proposition 4.2, f would have a periodic point with period 2 (since $I_{p-2} \cap I_{p-1} = \varnothing$). This contradiction shows that p cannot be the smallest even period. So one can consider a periodic point with smaller even period and restart the process.

This yields the desired result. □

We proceed with the proof of the theorem.

Case 1. $p = 2^k$ and $q = 2^l \prec p$ with $l < k$.

Take $l > 0$. If x is a periodic point of f with period p, then it is a periodic point of $f^{q/2}$ with period 2^{k-l+1}. Since $k - l + 1 \geq 2$, it follows from Lemma 4.6 that $f^{q/2}$ has a periodic point y with period 2, which is a periodic point of f with period q.

Now take $l = 0$. It follows from Lemma 4.6 that f has a periodic point with period 2. It determines an interval I_1 in I whose endpoints are permuted by f. Since f is continuous, it must have a fixed point in I_1.

Case 2. $p = 2^k r$ and $q = 2^k s \prec p$ with $r > 1$ odd minimal and s even.

We note that r is the smallest odd period of the periodic points of f^{2^k}. It follows from Lemma 4.5 that there exists a cycle of length s. More precisely, when $s < r$, take

$$I_{r-1} \to I_{r-s} \to \cdots \to I_{r-2} \to I_{r-1}$$

and when $s \geq r$, take

$$I_1 \to I_2 \to \cdots \to I_{r-1} \to I_1 \to I_1 \to \cdots \to I_1.$$

It follows from Proposition 4.2 that f^{2^k} has a periodic point with period s, which is a periodic point of f with period $2^k s = q$.

Case 3. $p = 2^k r$ and $q = 2^l \prec p$ with $r > 1$ odd minimal and $l \leq k$.

Taking $s = 2$ in Case 2, we obtain a periodic point of f with period $2^k s = 2^{k+1}$. It follows from Case 1 that f has a periodic point with period 2^l for each $l \leq k$.

Case 4. $p = 2^k r$ and $q = 2^k s \prec p$ with $r > 1$ odd minimal and $s > r$ odd.

Again, r is the smallest odd period of the periodic points of f^{2^k}. By Lemma 4.5, we obtain the cycle of length s given by

$$I_1 \to I_2 \to \cdots \to I_{r-1} \to I_1 \to I_1 \to \cdots \to I_1.$$

It follows from Proposition 4.2 that f^{2^k} has a periodic point x with period s. If x is a periodic point of f with period $2^k s$, then the proof is complete. Otherwise, x has period $2^l s$ for some $l < k$. Take $\bar{p} = 2^l s$ and $\bar{q} = 2^l \bar{s} = q$, where $\bar{s} = 2^{k-l} s$. Since $\bar{s}$ is even, it follows from Case 2 that there exists a periodic point of f with period $\bar{q} = q$.   □

4.4 The Poincaré–Bendixson Theorem

In this section we establish one of the most important results of the qualitative theory of differential equations in the plane: the Poincaré–Bendixson theorem.

Given a C^1 function $f \colon \mathbb{R}^2 \to \mathbb{R}^2$, consider the initial value problem

$$\begin{cases} x' = f(x), \\ x(0) = x_0 \end{cases} \tag{4.37}$$

for each $x_0 \in \mathbb{R}^2$. We assume that the unique solution $x(t, x_0)$ of (4.37) is defined for $t \in \mathbb{R}$. It follows from Proposition 2.3 that the family of maps $\varphi_t \colon \mathbb{R}^2 \to \mathbb{R}^2$ defined for each $t \in \mathbb{R}$ by $\varphi_t(x_0) = x(t, x_0)$ is a flow.

Now we establish the Poincaré–Bendixson theorem. We recall that a point $x \in \mathbb{R}^2$ with $f(x) = 0$ is called a *critical point* of f.

Theorem 4.10 (Poincaré–Bendixson) *Let $f \colon \mathbb{R}^2 \to \mathbb{R}^2$ be a C^1 function. For the flow φ_t determined by the equation $x' = f(x)$, if the positive semiorbit $\gamma^+(x)$ of a point $x \in \mathbb{R}^2$ is bounded and $\omega(x)$ contains no critical points, then $\omega(x)$ is a periodic orbit.*

Proof Since the positive semiorbit $\gamma^+(x)$ is bounded, it follows from Proposition 3.6 that $\omega(x)$ is nonempty. Take a point $p \in \omega(x)$. Since $\omega(x)$ is contained in the closure of $\gamma^+(x)$, it follows from the first property in Proposition 3.6 that $\omega(p)$ is nonempty and it follows from the second property in Proposition 3.5 that $\omega(p) \subset \omega(x)$. Now take a point $q \in \omega(p)$. By hypothesis, q is not a critical point and thus, there exists a line segment L containing q that is a transversal to f (see Exercise 3.5). Since $q \in \omega(p)$, it follows from the first property in Proposition 3.5 that there exists a sequence $t_k \nearrow +\infty$ in $\mathbb{R}^+$ such that $\varphi_{t_k}(p) \to q$ when $k \to \infty$. One can also assume that $\varphi_{t_k}(p) \in L$ for $k \in \mathbb{N}$. On the other hand, since $p \in \omega(x)$, it follows from the second property in Proposition 3.5 that $\varphi_{t_k}(p) \in \omega(x)$ for $k \in \mathbb{N}$. Since $\varphi_{t_k}(p) \in \omega(x) \cap L$, it follows from Exercise 3.5 that

$$\varphi_{t_k}(p) = \varphi_{t_l}(p) = q$$

for $k, l \in \mathbb{N}$. This implies that $\gamma(p) \subset \omega(x)$ is a periodic orbit.

Now we show that $\omega(x) = \gamma(p)$. Let us assume that $\omega(x) \setminus \gamma(p) \neq \varnothing$. Since $\omega(x)$ is connected (by Proposition 3.6), in each open neighborhood of $\gamma(p)$ there exist points of $\omega(x)$ that are not in $\gamma(p)$. Moreover, any sufficiently small open neighborhood of $\gamma(p)$ contains critical points. Thus, there exists a transversal L' to f containing one of these points, which is in $\omega(x)$, and a point of $\gamma(p)$. This shows that $\omega(x) \cap L'$ contains at least two points since $\gamma(p) \subset \omega(x)$, which contradicts Exercise 3.5. Thus, $\omega(x) = \gamma(p)$ and the ω-limit set of x is a periodic orbit. □

The following example is an application of Theorem 4.10.

Example 4.8 Consider the differential equation

$$\begin{cases} x' = x(3 - 2y - x^2 - y^2) - y, \\ y' = y(3 - 2y - x^2 - y^2) + x \end{cases} \tag{4.38}$$

that in polar coordinates takes the form

$$\begin{cases} r' = r(3 - 2r\sin\theta - r^2), \\ \theta' = 1. \end{cases}$$

For any sufficiently small r, we have

$$r' = r(3 - 2r\sin\theta - r^2) \geq r(3 - 2r - r^2) > 0 \tag{4.39}$$

and for any sufficiently large r, we have

$$r' = r(3 - 2r\sin\theta - r^2) \leq r(3 + 2r - r^2) < 0. \tag{4.40}$$

Since the origin is the only critical point, for any $r_2 > r_1 > 0$ there are no critical points in the ring

$$D = \{x \in \mathbb{R}^2 : r_1 < \|x\| < r_2\}.$$

Moreover, provided that r_1 is sufficiently small and r_2 is sufficiently large, it follows from (4.39) and (4.40) that any positive semiorbit $\gamma^+(x)$ of a point $x \in D$ is contained in D. By Theorem 4.10, the set $\omega(x) \subset D$ is a periodic orbit for each $x \in D$. In particular, the flow determined by Eq. (4.38) has at least one periodic orbit in the set D.

We have an analogous result for bounded negative semiorbits.

Theorem 4.11 *Let $f: \mathbb{R}^2 \to \mathbb{R}^2$ be a C^1 function. For the flow φ_t determined by the equation $x' = f(x)$, if the negative semiorbit $\gamma^-(x)$ of a point $x \in \mathbb{R}^2$ is bounded and $\alpha(x)$ contains no critical points, then $\alpha(x)$ is a periodic orbit.*

4.5 Exercises

Exercise 4.1 Find the fixed points of the map f in (4.4) for $|\beta| < 1/(2\pi)$.

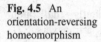

Fig. 4.5 An orientation-reversing homeomorphism

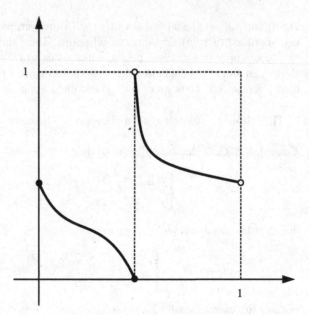

Exercise 4.2 Determine whether the map f in (4.4) has dense orbits for $|\beta| < 1/(2\pi)$.

Exercise 4.3 A homeomorphism $f\colon S^1 \to S^1$ is said to be *orientation-reversing* if at least one of its lifts is a decreasing function (see Fig. 4.5). Show that f is a orientation-reversing homeomorphism if and only if all its lifts are decreasing functions.

Exercise 4.4 Show that if $f, g\colon S^1 \to S^1$ are homeomorphism with lifts F and G, respectively, then $F \circ G$ is a lift of $f \circ g$.

Exercise 4.5 Show that the composition of two orientation-preserving homeomorphisms of the circle is still an orientation-preserving homeomorphism.

Exercise 4.6 Show that the composition of two orientation-reversing homeomorphisms of the circle is an orientation-preserving homeomorphism.

Exercise 4.7 Show that if $f\colon S^1 \to S^1$ is an orientation-preserving homeomorphism, then $\rho(f^n) = n\rho(f) \bmod 1$ for $n \in \mathbb{N}$.

Exercise 4.8 Let $f, g\colon S^1 \to S^1$ be orientation-preserving homeomorphisms. Show that if $f \circ g = g \circ f$, then

$$\rho(f \circ g) = \rho(f) + \rho(g) \bmod 1.$$

Exercise 4.9 Let $f, g\colon S^1 \to S^1$ be orientation-preserving homeomorphisms. Show that if f and g are topologically conjugate, then $\rho(f) = \rho(g)$.

Exercise 4.10 Show that each orientation-reversing homeomorphism of the circle has exactly two fixed points.

Exercise 4.11 Let $f\colon S^1 \to S^1$ be an orientation-preserving C^2 diffeomorphism. Show that if $\rho(f) \in \mathbb{R} \setminus \mathbb{Q}$, then f is topologically conjugate to the rotation $R_{\rho(f)}$.

Exercise 4.12 Show that any monotonous function $\varphi\colon S^1 \to \mathbb{R}$ has bounded variation.

Exercise 4.13 Given a function $\varphi\colon S^1 \to \mathbb{R}$, show that if there exists $L > 0$ such that

$$\left|\varphi(x) - \varphi(y)\right| \le L|x - y|$$

for $x, y \in S^1$, then φ has bounded variation.

Exercise 4.14 Consider the continuous function $\varphi\colon S^1 \to \mathbb{R}$ defined by

$$\varphi(x) = \begin{cases} x \sin(\pi/x) & \text{if } x \ne 0, \\ 0 & \text{if } x = 0. \end{cases}$$

1. Given $n \in \mathbb{N}$, show that

$$\sum_{i=1}^{n-1} \left|\varphi(x_i) - \varphi(x_{i-1})\right| = 4 \sum_{i=1}^{n} \frac{1}{2i + 1}$$

for the disjoint open intervals (x_{i-1}, x_i) with

$$x_0 = 0, \qquad x_1 = \frac{2}{2n+1}, \qquad x_2 = \frac{2}{2n-1}, \qquad \ldots, \qquad x_{n-1} = \frac{2}{5},$$

$$x_n = \frac{2}{3}, \qquad x_{n+1} = 1.$$

2. Conclude that φ does not have bounded variation.

Exercise 4.15 Determine whether there exists a homeomorphism f of the circle such that $f \circ E_2 = E_3 \circ f$.

Exercise 4.16 Given $\alpha \in \mathbb{R}$, determine whether there exists a homeomorphism f of the circle such that $f \circ R_\alpha = R_{-\alpha} \circ f$.

Exercise 4.17 Verify that the flow determined by the equation

$$\begin{cases} x' = y + x^3, \\ y' = -1 - x^4 \end{cases}$$

has no periodic orbits contained in the first quadrant.

Exercise 4.18 Verify that the flow determined by the equation

$$\begin{cases} x' = x(2 - x - x^2 - 2y^2) - y, \\ y' = y(2 - x - x^2 - 2y^2) + x \end{cases}$$

has a periodic orbit.

Exercise 4.19 Determine whether the flow defined by the equation

$$\begin{cases} r' = r(1 + \cos\theta), \\ \theta' = 1 \end{cases}$$

has a periodic orbit.

Exercise 4.20 Determine whether there exists a differential equation in $\mathbb{R}^2$ determining a flow with a dense orbit.

Chapter 5
Hyperbolic Dynamics I

This chapter is an introduction to hyperbolic dynamics. We first introduce the notion of a hyperbolic set. In particular, we describe the Smale horseshoe and some of its modifications. We also establish the continuity of the stable and unstable spaces on the base point. We then consider the characterization of a hyperbolic set in terms of invariant families of cones. In particular, this allows us to describe some stability properties of hyperbolic sets under sufficiently small perturbations. The prerequisites from the theory of smooth manifolds are fully recalled in Sect. 5.1.

5.1 Smooth Manifolds

In this section we recall some basic notions of the theory of smooth manifolds.

Definition 5.1 A set M is said to admit a *differentiable structure* of dimension $n \in \mathbb{N}$ if there exist injective maps $\varphi_i : U_i \to M$ in open sets $U_i \subset \mathbb{R}^n$ for $i \in I$ such that:

1. $\bigcup_{i \in I} \varphi_i(U_i) = M$;
2. for any $i, j \in I$ such that

$$V = \varphi_i(U_i) \cap \varphi_j(U_j) \neq \varnothing,$$

the preimages $\varphi_i^{-1}(V)$ and $\varphi_j^{-1}(V)$ are open and the map $\varphi_j^{-1} \circ \varphi_i$ is of class C^1.

Each map $\varphi_i : U_i \to M$ is called a *chart* or a *coordinate system*. Given a differentiable structure on M, we consider the topology on M formed by the sets $A \subset M$ such that $\varphi_i^{-1} A \subset \mathbb{R}^n$ is open for every $i \in I$.

Definition 5.2 A set M is said to be a *(smooth) manifold* of dimension n if it admits a differentiable structure of dimension n and is a Hausdorff topological space with countable basis.

L. Barreira, C. Valls, *Dynamical Systems*, Universitext, DOI 10.1007/978-1-4471-4835-7_5,   87
© Springer-Verlag London 2013

We recall that a topological space is said to be Hausdorff if any distinct points have disjoint open neighborhoods, and that it is said to have a countable basis if there exists a countable family of open sets such that each open set can be written as a union of elements of this family.

Example 5.1 Let $\varphi: U \to \mathbb{R}^m$ be a function of class C^1 in an open set $U \subset \mathbb{R}^n$. Then the graph

$$M = \{(x, \varphi(x)) : x \in U\} \subset \mathbb{R}^n \times \mathbb{R}^m$$

is a manifold of dimension n. A differentiable structure is given by the single map $\psi: U \to \mathbb{R}^n \times \mathbb{R}^m$ defined by $\psi(x) = (x, \varphi(x))$.

Example 5.2 The set

$$\mathbb{T} = \{(x, y) \in \mathbb{R}^2 : x^2 + y^2 = 1\}$$

is a manifold of dimension 1. A differentiable structure is given by the maps

$$\varphi_i: (-1, 1) \to \mathbb{T}, \quad i = 1, 2, 3, 4$$

defined by

$$\varphi_1(x) = (x, \sqrt{1 - x^2}), \qquad \varphi_2(x) = (x, -\sqrt{1 - x^2}),$$
$$\varphi_3(x) = (\sqrt{1 - x^2}, x), \qquad \varphi_4(x) = (-\sqrt{1 - x^2}, x). \tag{5.1}$$

We note that $\mathbb{T}$ can be identified with S^1. In particular, the map $\chi: S^1 \to \mathbb{T}$ defined by

$$\chi(x) = (\cos(2\pi x), \sin(2\pi x))$$

is a homeomorphism.

Example 5.3 The torus $\mathbb{T}^n = S^1$ is a manifold of dimension n. A differentiable structure is given by the maps $\psi: (-1, 1)^n \to \mathbb{T}^n$ defined by

$$\psi(x_1, \ldots, x_n) = ((\chi^{-1} \circ \psi_1)(x_1), \ldots, (\chi^{-1} \circ \psi_n)(x_n)),$$

where each ψ_i is any of the functions $\varphi_1, \varphi_2, \varphi_3$ and φ_4 in (5.1).

Now we introduce the notion of a differentiable map.

Definition 5.3 A map $f: M \to N$ between manifolds is said to be *differentiable* at a point $x \in M$ if there exist charts $\varphi: U \to M$ and $\psi: V \to N$ such that:

1. $x \in \varphi(U)$ and $f(\varphi(U)) \subset \psi(V)$;
2. $\psi^{-1} \circ f \circ \varphi$ is differentiable at $\varphi^{-1}(x)$.

Moreover, f is said to be of class C^k in an open set $W \subset M$ if all maps $\psi^{-1} \circ f \circ \varphi$ are of class C^k in $\varphi^{-1}(W)$.

We also recall the notion of a tangent vector. Let M be a manifold of dimension n and let D_x be the set of all functions $g \colon M \to \mathbb{R}$ that are differentiable at $x \in M$.

Definition 5.4 The *tangent vector* to a differentiable path $\alpha \colon (-\varepsilon, \varepsilon) \to M$ with $\alpha(0) = x$ at $t = 0$ is the function $v_\alpha \colon D_x \to \mathbb{R}$ defined by

$$v_\alpha(g) = \frac{d(g \circ \alpha)}{dt}\bigg|_{t=0}.$$

We also say that v_α is a *tangent vector* at x.

One can show that the set $T_x M$ of all tangent vectors at x is a vector space of dimension n. It is called the *tangent space* of M at x. Moreover, the set

$$TM = \{(x, v) : x \in M, \ v \in T_x M\}$$

is a manifold of dimension $2n$, called the *tangent bundle* of M. A differentiable structure can be obtained as follows. Let $\varphi \colon U \to M$ be a chart and let $(x_1, \ldots, x_n)$ be the coordinates in U. Consider the differentiable paths $\alpha_i \colon (-\varepsilon, \varepsilon) \to M$ for $i = 1, \ldots, n$ defined by $\alpha_i(t) = \varphi(t e_i)$, where $(e_1, \ldots, e_n)$ is the standard basis of $\mathbb{R}^n$. The tangent vector to the path α_i at $t = 0$ is denoted by $\partial/\partial x_i$. One can show that $(\partial/\partial x_1, \ldots, \partial/\partial x_n)$ is a basis of the tangent space $T_{\varphi(0)} M$ and that a differentiable structure on TM is given by the maps $\psi \colon U \times \mathbb{R}^n \to TM$ defined by

$$\psi(x_1, \ldots, x_n, y_1, \ldots, y_n) = \left(\varphi(x_1, \ldots, x_n), \sum_{i=1}^{n} y_i \frac{\partial}{\partial x_i} \right).$$

5.2 Hyperbolic Sets

In this section we introduce the notion of a hyperbolic set. We also give some examples of hyperbolic sets and we establish the continuity of the stable and unstable spaces on the base point.

5.2.1 Basic Notions

Let $f \colon M \to M$ be a C^1 diffeomorphism of a manifold M (this means that f is an invertible C^1 map whose inverse is also of class C^1). For each $x \in M$, we define a linear transformation

$$d_x f \colon T_x M \to T_{f(x)} M$$

between the tangent spaces $T_x M$ and $T_{f(x)} M$ by

$$d_x f v = v_{f \circ \alpha}$$

for any differentiable path $\alpha \colon (-\varepsilon, \varepsilon) \to M$ such that $\alpha(0) = x$ and $v_\alpha = v$ (one can show that the definition does not depend on the path α).

We always assume that M is a Riemannian manifold. This means that each tangent space $T_x M$ is equipped with an inner product $\langle \cdot, \cdot \rangle_x$ such that the map $TM \ni (x, v) \mapsto \langle v, v \rangle'_x$ is differentiable. It induces the norm

$$\|v\|_x = \langle v, v \rangle_x^{1/2} \quad \text{for } v \in T_x M.$$

For simplicity of notation, we always write $\langle \cdot, \cdot \rangle$ and $\|\cdot\|$, without indicating the dependence on x (which can easily be deduced from the context).

Now we introduce the notion of a hyperbolic set.

Definition 5.5 A compact f-invariant set $\Lambda \subset M$ is said to be a *hyperbolic set* for f if there exist $\lambda \in (0, 1)$, $c > 0$, and a decomposition

$$T_x M = E^s(x) \oplus E^u(x) \tag{5.2}$$

for each $x \in \Lambda$ such that:

1.
$$d_x f E^s(x) = E^s(f(x)) \quad \text{and} \quad d_x f E^u(x) = E^u(f(x)); \tag{5.3}$$

2. if $v \in E^s(x)$ and $n \in \mathbb{N}$, then

$$\left\| d_x f^n v \right\| \leq c \lambda^n \|v\|;$$

3. if $v \in E^u(x)$ and $n \in \mathbb{N}$, then

$$\left\| d_x f^{-n} v \right\| \leq c \lambda^n \|v\|. \tag{5.4}$$

The linear spaces $E^s(x)$ and $E^u(x)$ are called, respectively, the *stable* and *unstable* *spaces* at the point x.

We first consider the particular case of the fixed points.

Example 5.4 Given $a \in (0, 1)$ and $b > 1$, let $f \colon \mathbb{R}^2 \to \mathbb{R}^2$ be the linear transformation defined by

$$f(x, y) = (ax, by).$$

We have $f(0) = 0$ and hence, the origin is a fixed point. We also consider the decomposition $\mathbb{R}^2 = E^s \oplus E^u$, where E^s and E^u are, respectively, the horizontal and vertical axes. For the linear transformation $A = d_0 f = f$, we have:

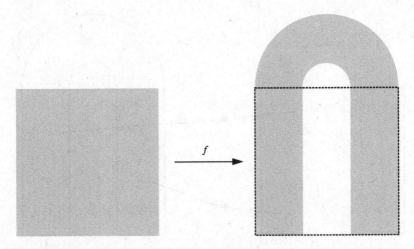

Fig. 5.1 A diffeomorphism f on an open neighborhood of the square Q

1. $AE^s = E^s$ and $AE^u = E^u$;
2. $\|Av\| \le a\|v\|$ for $v \in E^s$;
3. $\|A^{-1}v\| \le b^{-1}\|v\|$ for $v \in E^u$.

This shows that $\{0\} \subset \mathbb{R}^2$ is a hyperbolic set for the diffeomorphism f, taking $\lambda = \max\{a, b^{-1}\}$ and $c = 1$.

More generally, one can show that if $x \in M$ is a fixed point of f, then $\{x\}$ is a hyperbolic set for f if and only if the linear transformation $d_x f : T_x M \to T_x M$ has no eigenvalues with modulus 1 (see Exercise 5.14).

5.2.2 Smale Horseshoe

In this section we consider other examples of hyperbolic sets. More precisely, we consider the Smale horseshoe and some of its modifications (see also Sect. 7.4).

Let f be a diffeomorphism on an open neighborhood of the square $Q = [0, 1]^2$ with the behavior shown in Fig. 5.1. Consider the horizontal strips

$$H_1 = [0, 1] \times [0, a] \quad \text{and} \quad H_2 = [0, 1] \times [1 - a, 1]$$

and the vertical strips

$$V_1 = [0, a] \times [0, 1] \quad \text{and} \quad V_2 = [1 - a, 1] \times [0, 1], \qquad (5.5)$$

for some constant $a \in (0, 1/2)$. We assume that

$$f(H_1) = V_1 \quad \text{and} \quad f(H_2) = V_2 \qquad (5.6)$$

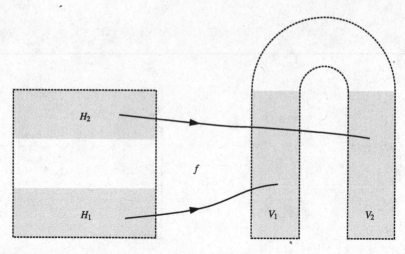

Fig. 5.2 Horizontal and vertical strips

(see Fig. 5.2), which yields the identity

$$Q \cap f(Q) = V_1 \cup V_2. \tag{5.7}$$

We also assume that the restrictions $f|H_1$ are $f|H_2$ are affine, with

$$f(x, y) = \begin{cases} (ax, by) & \text{if } (x, y) \in H_1, \\ (-ax + 1, -by + b) & \text{if } (x, y) \in H_2, \end{cases} \tag{5.8}$$

where $b = 1/a$. We shall see that the construction of the Smale horseshoe only depends on the restriction $f|(H_1 \cup H_2)$.

Now we consider the diffeomorphism f^{-1}. By (5.6), we have

$$f^{-1}(V_1) = H_1 \quad \text{and} \quad f^{-1}(V_2) = H_2$$

and thus, it follows from (5.7) that

$$f^{-1}(Q) \cap Q = f^{-1}(V_1) \cup f^{-1}(V_2) = H_1 \cup H_2. \tag{5.9}$$

Combining (5.7) and (5.9), we conclude that

$$\bigcap_{k=-1}^{1} f^n(Q) = (H_1 \cup H_2) \cap (V_1 \cap V_2)$$

is the union of 4 squares of size a (see Fig. 5.3 for an example with $a = 1/3$).

Fig. 5.3 The intersection
$f^{-1}(Q) \cap Q \cap f(Q)$

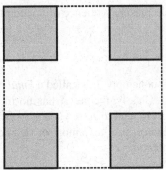

Fig. 5.4 The intersection
$\Lambda_2 = \bigcap_{k=-2}^{2} f^k(Q)$

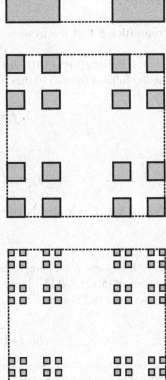

Fig. 5.5 The intersection
$\Lambda_3 = \bigcap_{k=-3}^{3} f^k(Q)$

Iterating this procedure, that is, considering successively the images $f^n(Q)$ and the preimages $f^{-n}(Q)$, we find that the intersection

$$\Lambda_n = \bigcap_{k=-n}^{n} f^k(Q)$$

is the union of 4^n squares of size a^n. For example, for $n = 2$ and $n = 3$ we obtain, respectively, the sets in Figs. 5.4 and 5.5 (again for $a = 1/3$). Since Λ_n is a decreasing sequence of nonempty closed sets, the compact set

$$\Lambda = \bigcap_{n\in\mathbb{N}} \Lambda_n = \bigcap_{k\in\mathbb{Z}} f^k(Q) \tag{5.10}$$

is nonempty. It is called a *Smale horseshoe* (for f).

Clearly, the set Λ has no interior points since the diameters of the 4^n squares in Λ_n tend to zero when $n \to \infty$. One can also verify that Λ has no isolated points. Hence, it is a Cantor set (it is closed and has neither interior points nor isolated points).

Proposition 5.1 Λ *is a hyperbolic set for the diffeomorphism* f.

Proof It follows from (5.10) that Λ is f-invariant, that is, $f^{-1}\Lambda = \Lambda$. On the other hand, it follows from (5.8) that

$$d_x f = \begin{pmatrix} a & 0 \\ 0 & b \end{pmatrix} \quad \text{for } x \in H_1 \tag{5.11}$$

and

$$d_x f = \begin{pmatrix} -a & 0 \\ 0 & -b \end{pmatrix} \quad \text{for } x \in H_2. \tag{5.12}$$

For each $x \in \Lambda$, consider the decomposition

$$\mathbb{R}^2 = E^s(x) \oplus E^u(x),$$

where $E^s(x)$ and $E^u(x)$ are, respectively, the horizontal and vertical axes. Since the matrices in (5.11) and (5.12) are diagonal, the identities in (5.3) hold. Moreover, it follows from (5.11) and (5.12) that

$$\|d_x f v\| = \begin{cases} a\|v\| & \text{if } v \in E^s(x), \\ b\|v\| & \text{if } v \in E^u(x). \end{cases}$$

Hence, one can take $\lambda = a$ and $c = 1$ in the definition of a hyperbolic set. $\square$

One can also consider other types of constructions. For example, let g be a diffeomorphism on an open neighborhood of the square Q with the behavior shown in Fig. 5.6. We assume that the identities in (5.6) hold and that

$$g(x, y) = \begin{cases} (x/3, 3y) & \text{if } (x, y) \in H_1, \\ (x/3 + 2/3, 3y - 2) & \text{if } (x, y) \in H_2. \end{cases}$$

Then the compact g-invariant set

$$\Lambda_g = \bigcap_{n\in\mathbb{Z}} g^n(Q)$$

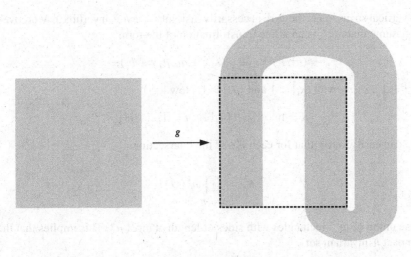

Fig. 5.6 A diffeomorphism g on an open neighborhood of Q

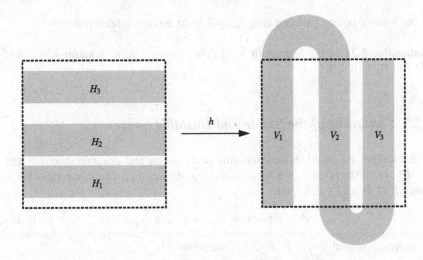

Fig. 5.7 A diffeomorphism h on an open neighborhood of Q

coincides with the set Λ in (5.10). One can also show that it is a hyperbolic set.

Proposition 5.2 Λ_g *is a hyperbolic set for the diffeomorphism* g.

We mention still another modification of the initial construction of the Smale horseshoe. Let h be a diffeomorphism on an open neighborhood of the square Q such that $Q \cap h(Q)$ has a finite number of connected components. More precisely, consider pairwise disjoint closed horizontal strips $H_1, \ldots, H_m \subset Q$ (see Fig. 5.7 for an example for $m = 3$). We assume that the images $V_i = h(H_i)$, for $i = 1, \ldots, m$, are vertical strips in Q (they are necessarily disjoint since h is invertible). Moreover, we assume that $h|H_i$ is an affine transformation of the form

$$(h|H_i)(x, y) = (\lambda_i x + a_i, \mu_i y + b_i),$$

for $i = 1, \ldots, m$, with $|\lambda_i| < 1$ and $|\mu_i| > 1$. Now let

$$\mu = \max\{|\lambda_i|, |\mu_i|^{-1} : i = 1, \ldots, m\}.$$

One can easily verify that for each $n \in \mathbb{N}$ the intersection

$$\Lambda_n^h = \bigcap_{k=-n}^{n} h^k(Q)$$

is the union of m^{2n} rectangles with sides of length at most μ^n. This implies that the compact h-invariant set

$$\Lambda_h = \bigcap_{n \in \mathbb{Z}} h^n(Q)$$

has no interior points. One can also verify that Λ_h has no isolated points.

Proposition 5.3 Λ_h *is a hyperbolic set for the diffeomorphism h, taking $\lambda = \mu$ and $c = 1$ in Definition 5.5.*

5.2.3 Continuity of the Stable and Unstable Spaces

In this section we establish the continuity of the stable and unstable spaces $E^s(x)$ and $E^u(x)$ on the point x. We first introduce a distance between subspaces of $\mathbb{R}^p$. Given $E \subset \mathbb{R}^p$ and $v \in \mathbb{R}^p$, let

$$d(v, E) = \min\{\|v - w\| : w \in E\}. \tag{5.13}$$

Moreover, given subspaces $E, F \subset \mathbb{R}^p$, we define

$$d(E, F) = \max\left\{\max_{v \in E, \|v\|=1} d(v, F), \max_{w \in F, \|w\|=1} d(w, E)\right\}.$$

Example 5.5 If $E, F \subset \mathbb{R}^2$ are subspaces of dimension 1, then $d(E, F) = \sin\alpha$, where $\alpha \in [0, \pi/2]$ is the angle between E and F. Indeed, in this case, we have

$$\max_{v \in E, \|v\|=1} d(v, F) = d(v_E, F) \quad \text{and} \quad \max_{v \in F, \|w\|=1} d(w, E) = d(v_F, E),$$

where $v_E \in E$ and $v_F \in F$ are any vectors with norm 1. These numbers coincide and hence,

$$d(E, F) = d(v_E, F) = d(v_F, E) = \sin\alpha.$$

Now we consider the stable and unstable spaces $E^s(x)$ and $E^u(x)$ for x in a hyperbolic set $\Lambda \subset \mathbb{R}^p$.

Theorem 5.1 *If $\Lambda \subset \mathbb{R}^p$ is a hyperbolic set, then the spaces $E^s(x)$ and $E^u(x)$ vary continuously with $x \in \Lambda$, that is, if $x_m \to x$ when $m \to \infty$, with $x_m, x \in \Lambda$ for each $m \in \mathbb{N}$, then*

$$d\big(E^s(x_m), E^s(x)\big) \to 0 \quad \text{when } m \to \infty$$

and

$$d\big(E^u(x_m), E^u(x)\big) \to 0 \quad \text{when } m \to \infty.$$

Proof Let $(x_m)_{m \in \mathbb{N}}$ be a sequence as in the statement of the theorem.

Lemma 5.1 *Any sublimit of a sequence $v_m \in E^s(x_m) \subset \mathbb{R}^p$ with $\|v_m\| = 1$ is in $E^s(x)$.*

Proof Since the closed unit sphere of $\mathbb{R}^p$ is compact, the sequence v_m has sublimits. On the other hand, since $v_m \in E^s(x_m)$, we have

$$\big\| d_{x_m} f^n v_m \big\| \le c\lambda^n \|v_m\|$$

for $m, n \in \mathbb{N}$. Letting $m \to \infty$, we obtain

$$\big\| d_x f^n v \big\| \le c\lambda^n \|v\|$$

for $n \in \mathbb{N}$, where v is any sublimit of the sequence v_m. Finally, it follows from (5.4) that v has no component in $E^u(x)$ and thus, $v \in E^s(x)$. $\qquad\square$

Lemma 5.2 *There exists an $m \in \mathbb{N}$ such that*

$$\dim E^s(x_p) = \dim E^s(x_q) \quad \text{and} \quad \dim E^u(x_p) = \dim E^s(x_q)$$

for any $p, q > m$.

Proof Since the dimensions $\dim E^s(x_m)$ and $\dim E^u(x_m)$ can only take finitely many values, there exists a subsequence y_m of x_m such that the numbers $\dim E^s(y_m)$ and $\dim E^u(y_m)$ are independent of m. Now let

$$v_{1m}, \ldots, v_{km} \in E^s(y_m) \subset \mathbb{R}^p$$

be an orthonormal basis of $E^s(y_m)$, where $k = \dim E^s(y_m)$ (that by hypothesis is independent of m). Since the closed unit sphere of $\mathbb{R}^p$ is compact, the sequence $(v_{1m}, \ldots, v_{km})$ has sublimits. Moreover, each sublimit $(v_1, \ldots, v_k)$ is still an orthonormal set. It follows from Lemma 5.1 that $v_1, \ldots, v_k \in E^s(x)$ and thus $\dim E^s(x) \ge k$ (since $(v_1, \ldots, v_k)$ is an orthonormal set). Proceeding analogously for the unstable spaces, we obtain $\dim E^u(x) \geq \dim M - k$. It follows from (5.2) that

$$\dim E^s(x) = k \quad \text{and} \quad \dim E^u(x) = \dim M - k. \tag{5.14}$$

In particular, the vectors $v_1, \ldots, v_k$ generate $E^s(x)$. We also note that each vector $v \in E^s(x)$ with norm $\|v\| = 1$ is a sublimit of some sequence $v_m \in E^s(y_m)$ with $\|v_m\| = 1$. Indeed, writing $v = \sum_{i=1}^{k} \alpha_i v_i$ with $\sum_{i=1}^{k} \alpha_i^2 = 1$, one can take

$$v_m = \sum_{i=1}^{k} \alpha_i v_{im} \Big/ \left\| \sum_{i=1}^{k} \alpha_i v_{im} \right\|.$$

If z_m is another subsequence of x_m such that the dimensions $\dim E^s(z_m)$ and $\dim E^u(z_m)$ are independent of m, respectively, with values l and $\dim M - l$, then we also have

$$\dim E^s(x) = l \quad \text{and} \quad \dim E^u(x) = \dim M - l.$$

Comparing with (5.14), we conclude that $l = k$. This shows that $\dim E^s(x_m)$ and $\dim E^u(x_m)$ are constant for any sufficiently large m. ☐

Now we estimate the distance $d(E^s(x_m), E^s(x))$.

Lemma 5.3 *Given $\delta > 0$, there exists a $p \in \mathbb{N}$ such that*

$$\max_{w \in E^s(x_m), \|w\|=1} d(w, E^s(x)) < \delta \quad \text{for } m > p. \tag{5.15}$$

Proof We note that given $\varepsilon > 0$ and a sequence $w_m \in E^s(x_m)$ with $\|w_m\| = 1$, we have $d(w_m, E^s(x)) < \varepsilon$ for any sufficiently large m. Otherwise, there would exist a subsequence w_{k_m} such that

$$d(w_{k_m}, E^s(x)) \geq \varepsilon \quad \text{for } m \in \mathbb{N}.$$

By (5.13), any sublimit w of the sequence w_{k_m} satisfies $d(w, E^s(x)) \geq \varepsilon$. But this is impossible since by Lemma 5.1, we have $w \in E^s(x)$.

Now we consider orthonormal bases $(v_{1m}, \ldots, v_{km})$ of $E^s(x_m)$ (for each sufficiently large m such that $\dim E^s(x_m) = k$). It follows from the former paragraph that there exist integers $p_1, \ldots, p_k \in \mathbb{N}$ such that

$$d(v_{im}, E^s(x)) < \varepsilon \quad \text{for } m > p_i. \tag{5.16}$$

We also take vectors $w_m \in E^s(x_m)$ with norm $\|w_m\| = 1$ and we write

$$w_m = \sum_{i=1}^{k} \alpha_{im} v_{im}$$

with $\sum_{i=1}^{k} \alpha_{im}^2 = 1$. By (5.16), for each $i = 1, \ldots, k$ and $m > p := \max\{p_1, \ldots, p_k\}$, there exists a $w_{im} \in E^s(x)$ such that $\|v_{im} - w_{im}\| < \varepsilon$. Then

$$d\big(w_m, E^s(x)\big) \le \left\| w_m - \sum_{i=1}^{k} \alpha_{im} w_{im} \right\|$$

$$\le \sum_{i=1}^{k} |\alpha_{im}| \cdot \|v_{im} - w_{im}\| < k\varepsilon$$

and hence,

$$\max_{w \in E^s(x_m), \|w\|=1} d\big(w, E^s(x)\big) < k\varepsilon \quad \text{for } m > p.$$

This establishes the desired result.   □

Lemma 5.4 *Given $\delta > 0$, there exists a $q \in \mathbb{N}$ such that*

$$\max_{v \in E^s(x), \|v\|=1} d\big(v, E^s(x_m)\big) < \delta \quad \text{for } m > q. \tag{5.17}$$

Proof Given $\varepsilon > 0$ and $v \in E^s(x)$, we show that $d(v, E^s(x_m)) < \varepsilon$ for any sufficiently large m. Otherwise, there would exist a sequence x_{k_m} such that

$$d\big(v, E^s(x_{k_m})\big) \ge \varepsilon \quad \text{for } m \in \mathbb{N}.$$

Now we consider a sequence $w_m \in E^s(x_{k_m})$ with $\|w_m\| = 1$ having v as a sublimit (we recall that each element of $E^s(x)$ is obtained as a sublimit of vectors w_m of this form). But this is impossible since then we would have $\|v - w_m\| \ge \varepsilon$ for $m \in \mathbb{N}$ and thus also $0 = \|v - v\| \ge \varepsilon$. Now we consider an orthonormal basis $v_1, \ldots, v_k$ of $E^s(x)$ and we take integers $q_1, \ldots, q_k \in \mathbb{N}$ such that

$$d\big(v_i, E^s(x_m)\big) < \varepsilon \quad \text{for } m > q_i.$$

For each i, there exists a $v_{im} \in E^s(x_m)$ with $\|v_i - v_{im}\| < \varepsilon$. Given $v \in E^s(x)$ with norm $\|v\| = 1$, we write $v = \sum_{i=1}^{k} \alpha_i v_i$ with $\sum_{i=1}^{k} \alpha_i^2 = 1$. Then

$$d\big(v, E^s(x_m)\big) \le \left\| v - \sum_{i=1}^{k} \alpha_i v_{im} \right\|$$

$$\le \sum_{i=1}^{k} |\alpha_i| \cdot \|v_i - v_{im}\| < k\varepsilon$$

for $m > q := \max\{q_1, \ldots, q_k\}$ and hence,

$$\max_{v \in E^s(x), \|v\|=1} d\big(v, E^s(x_m)\big) < k\varepsilon \quad \text{for } m > q.$$

This establishes the desired result.   □

Finally, it follows from (5.15) and (5.17) that

$$d\big(E^s(x_m), E^s(x)\big) < 2\delta \quad \text{for } m > \max\{p, q\}.$$

We obtain in a similar manner the corresponding result for the unstable spaces. □

Lemma 5.2 says that if $x_m \to x$ when $m \to \infty$, with $x_m, x \in \Lambda$ for each $m \in \mathbb{N}$, then $\dim E^s(x_m)$ and $\dim E^u(x_m)$ are constant for any sufficiently large m.

5.3 Hyperbolic Sets and Invariant Families of Cones

In this section we describe a characterization of hyperbolic sets in terms of invariant families of cones.

5.3.1 Formulation of the Result

Let $f : M \to M$ be a C^1 diffeomorphism and let $\Lambda \subset M$ be a compact f-invariant set. For each $x \in \Lambda$, we consider a decomposition

$$T_x M = F^s(x) \oplus F^u(x) \tag{5.18}$$

and an inner product $\langle \cdot, \cdot \rangle' = \langle \cdot, \cdot \rangle'_x$ in $T_x M$. We emphasize that this may not be the original inner product. We always assume that the dimensions $\dim F^s(x)$ and $\dim F^u(x)$ are independent of x. On the other hand, we do not require that

$$d_x f F^s(x) = F^s(f(x)) \quad \text{and} \quad d_x f F^u(x) = F^u(f(x))$$

for $x \in \Lambda$.

Definition 5.6 Given $\gamma \in (0, 1)$ and $x \in \Lambda$, we define the *cones*

$$C^s(x) = \big\{(v, w) \in F^s(x) \oplus F^u(x) : \|w\|' < \gamma \|v\|'\big\} \cup \{0\} \tag{5.19}$$

and

$$C^u(x) = \big\{(v, w) \in F^s(x) \oplus F^u(x) : \|v\|' < \gamma \|w\|'\big\} \cup \{0\} \tag{5.20}$$

(see Figs. 5.8 and 5.9).

The following result gives a characterization of a hyperbolic set in terms of cones.

Theorem 5.2 *Let $f : M \to M$ be a C^1 diffeomorphism and let $\Lambda \subset M$ be a compact f-invariant set. Then Λ is a hyperbolic set for f if and only if there exist a decomposition (5.18) and an inner product $\langle \cdot, \cdot \rangle'_x$ in $T_x M$, for each $x \in \Lambda$, and*

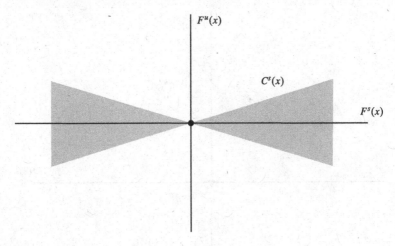

Fig. 5.8 The cone $C^s(x)$

constants $\mu, \gamma \in (0, 1)$ such that:

1. for any $x \in \Lambda$,

$$d_x f \overline{C^u(x)} \subset C^u(f(x)) \quad and \quad d_x f^{-1}\overline{C^s(x)} \subset C^s(f^{-1}(x)); \tag{5.21}$$

2. for any $x \in \Lambda$,

$$\|d_x f v\|' \geq \mu^{-1}\|v\|' \quad for\ v \in C^u(x) \tag{5.22}$$

and

$$\|d_x f^{-1}v\|' \geq \mu^{-1}\|v\|' \quad for\ v \in C^s(x). \tag{5.23}$$

Theorem 5.2 is an immediate consequence of Theorems 5.3 and 5.4 proven, respectively, in Sects. 5.3.2 and 5.3.3.

5.3.2 Existence of Invariant Families of Cones

In this section we show that any hyperbolic set has associated families of cones $C^s(x)$ and $C^u(x)$ with the properties in Theorem 5.2.

Theorem 5.3 *Let $f: M \to M$ be a C^1 diffeomorphism and let $\Lambda \subset M$ be a hyperbolic set for f. Then there exist an inner product $\langle \cdot, \cdot \rangle'_x$ in $T_x M$ varying continuously with $x \in \Lambda$ and constants $\mu, \gamma \in (0, 1)$ such that, taking*

$$F^s(x) = E^s(x) \quad and \quad F^u(x) = E^u(x) \tag{5.24}$$

in (5.19) and (5.20), the cones $C^s(x)$ and $C^u(x)$ satisfy properties (5.21), (5.22) and (5.23) for any $x \in \Lambda$.

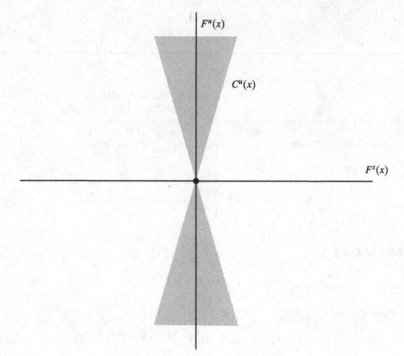

Fig. 5.9 The cone $C^u(x)$

Proof We divide the proof into steps.

Step 1. Construction of an inner product.

Take $m \in \mathbb{N}$ such that $c\lambda^m < 1$. Given $v, w \in E^s(x)$, we define

$$\langle v, w \rangle' = \sum_{n=0}^{m-1} \langle d_x f^n v, d_x f^n w \rangle.$$

For each $v \in E^s(x)$, we have

$$\left(\|d_x f v\|' \right)^2 = \sum_{n=0}^{m-1} \|d_x f^{n+1} v\|^2$$

$$= \sum_{n=0}^{m-1} \|d_x f^n v\|^2 - \|v\|^2 + \|d_x f^m v\|^2$$

$$\leq \left(\|v\|' \right)^2 - \left(1 - c^2 \lambda^{2m} \right) \|v\|^2.$$

On the other hand,

$$\left(\|v\|'\right)^2 \le \sum_{n=0}^{m-1} c^2 \lambda^{2n} \|v\|^2 \le c^2 m \|v\|^2 \tag{5.25}$$

and thus,

$$\|d_x f v\|' \le \tau \|v\|', \tag{5.26}$$

where

$$\tau = \sqrt{1 - \frac{1 - c^2 \lambda^{2m}}{c^2 m}} < 1.$$

Analogously, given $v, w \in E^u(x)$, we define

$$\langle v, w \rangle' = \sum_{n=0}^{m-1} \langle d_x f^{-n} v, d_x f^{-n} w \rangle.$$

One can verify that

$$\left\|d_x f^{-1} v\right\|' \le \tau \|v\|' \tag{5.27}$$

for $v \in E^u(x)$. Now we consider the inner product $\langle \cdot, \cdot \rangle = \langle \cdot, \cdot \rangle_x$ in $T_x M$ defined by

$$\langle v, w \rangle' = \langle v^s, w^s \rangle' + \langle v^u, w^u \rangle' \tag{5.28}$$

for each $v, w \in T_x M$, where

$$v = v^s + v^u \quad \text{and} \quad w = w^s + w^u,$$

with $v^s, w^s \in E^s(x)$ and $v^u, w^u \in E^u(x)$. We also take $F^s(x)$ and $F^u(x)$ as in (5.24), and we consider the cones $C^s(x)$ and $C^u(x)$ in (5.19) and (5.20), with the norm $\|\cdot\|'$ induced from the inner product in (5.28).

Step 2. Invariance of the families of cones.

Given $(v, w) \in \overline{C^u(x)}$, we have

$$\|v\|' \le \gamma \|w\|'$$

and it follows from (5.3) that

$$d_x f(v, w) = (d_x f v, d_x f w) \in E^s(f(x)) \oplus E^u(f(x)).$$

Using (5.26) and (5.27), we obtain

$$\|d_x f v\|' \le \tau \|v\|'$$
$$\le \tau \gamma \|w\|'$$
$$\le \tau^2 \gamma \|d_x f w\|'$$

and thus, $d_x f(v, w) \in C^u(f(x))$. Analogously, given $(v, w) \in \overline{C^s(x)}$, we have

$$\|w\|' \le \gamma \|v\|'$$

and thus,

$$\begin{aligned} \left\| d_x f^{-1} w \right\|' &\le \tau \|w\|' \\ &\le \tau \gamma \|v\|' \\ &\le \tau^2 \gamma \left\| d_x f^{-1} v \right\|'. \end{aligned}$$

This shows that $d_x f^{-1}(v, w) \in C^s(f^{-1}(x))$ and we obtain the inclusions in (5.21).

Step 3. Estimates inside the cones.

Given $(v, w) \in C^u(x)$, it follows from (5.26) and (5.27) that

$$\begin{aligned} \|d_x f(v, w)\|' &\ge \|d_x f w\|' - \|d_x f v\|' \\ &\ge \tau^{-1} \|w\|' - \tau \|v\|' \\ &\ge \tau^{-1} \|w\|' - \tau \gamma \|w\|'. \end{aligned}$$

Since

$$\|(v, w)\|' < (1 + \gamma)\|w\|',$$

we have

$$\|d_x f(v, w)\|' \ge \frac{\tau^{-1} - \tau \gamma}{1 + \gamma} \|(v, w)\|'.$$

Taking γ sufficiently small so that

$$\mu := \left(\frac{\tau^{-1} - \tau \gamma}{1 + \gamma} \right)^{-1} > 1,$$

we obtain property (5.22). Analogously, given $(v, w) \in C^s(x)$, it follows from (5.26) and (5.27) that

$$\begin{aligned} \left\| d_x f^{-1}(v, w) \right\|' &\ge \left\| d_x f^{-1} v \right\|' - \left\| d_x f^{-1} w \right\|' \\ &\ge \tau^{-1} \|v\|' - \tau \|w\|' \\ &\ge \frac{\tau^{-1} - \tau \gamma}{1 + \gamma} \|(v, w)\|' \\ &= \mu^{-1} \|(v, w)\|', \end{aligned}$$

which yields property (5.23). This completes the proof of the theorem.   □

5.3.3 Criterion for Hyperbolicity

The following result shows that the existence of cones $C^s(x)$ and $C^u(x)$ as in Theorem 5.2 guarantees that the compact f-invariant set Λ is a hyperbolic set for f.

Theorem 5.4 *Let $f \colon M \to M$ be a C^1 diffeomorphism and let $\Lambda \subset M$ be a compact f-invariant set. If there exist a decomposition (5.18) and an inner product $\langle \cdot, \cdot \rangle_x'$ in $T_x M$, for each $x \in \Lambda$, and constants $\mu, \gamma \in (0,1)$ such that the cones $C^s(x)$ and $C^u(x)$ satisfy properties (5.21), (5.22) and (5.23) for any $x \in \Lambda$, then Λ is a hyperbolic set for f, taking $\lambda = \mu$ and $c = 1$. Moreover, the stable and unstable spaces are given by*

$$E^s(x) = \bigcap_{n=0}^{\infty} d_{f^n(x)} \overline{C^s\big(f^n(x)\big)}$$

and

$$E^u(x) = \bigcap_{n=0}^{\infty} d_{f^{-n}(x)} \overline{C^u\big(f^{-n}(x)\big)}.$$

Proof We divide the proof into steps.

Step 1. Construction of invariant sets.

For each $x \in \Lambda$, we consider the sets

$$G^s(x) = \bigcap_{n=0}^{\infty} d_{f^n(x)} f^{-n} \overline{C^s\big(f^n(x)\big)}$$

and

$$G^u(x) = \bigcap_{n=0}^{\infty} d_{f^{-n}(x)} f^n \overline{C^u\big(f^{-n}(x)\big)}.$$

By (5.21), we have

$$G^s(x) \subset C^s(x) \quad \text{and} \quad G^u(x) \subset C^u(x), \tag{5.29}$$

and thus,

$$d_x f^{-1} G^s(x) \subset C^s\big(f^{-1}(x)\big) \quad \text{and} \quad d_x f G^u(x) \subset C^u(f(x)).$$

Writing $y = f^{-1}(x)$, we obtain

$$d_x f^{-1} G^s(x) = \overline{C^s(y)} \cap d_x f^{-1} G^s(x)$$

$$= \overline{C^s(y)} \cap \bigcap_{n=0}^{\infty} d_{f^n(x)} f^{-(n+1)} \overline{C^s\big(f^n(x)\big)}$$

$$= \overline{C^s(y)} \cap \bigcap_{n=0}^{\infty} d_{f^{n+1}(y)} f^{-(n+1)} \overline{C^s\left(f^{n+1}(y)\right)} = G^s(y) \qquad (5.30)$$

and analogously,

$$d_x f G^u(x) = G^u(f(x)). \qquad (5.31)$$

Step 2. Construction of stable and unstable spaces.

Since the dimensions $k = \dim F^s(x)$ and $l = \dim F^u(x)$ are independent of x, for each $m \in \mathbb{N}$ the sets

$$\bigcap_{n=0}^{m} d_{f^n(x)} f^{-n} \overline{C^s\left(f^n(x)\right)} = d_{f^m(x)} f^{-m} \overline{C^s\left(f^m(x)\right)}$$

and

$$\bigcap_{n=0}^{m} d_{f^{-n}(x)} f^n \overline{C^u\left(f^{-n}(x)\right)} = d_{f^{-m}(x)} f^m \overline{C^u\left(f^{-m}(x)\right)}$$

contain subspaces $E_m^s(x)$ and $E_m^u(x)$, respectively, of dimensions

$$\dim E_m^s(x) = k \quad \text{and} \quad \dim E_m^u(x) = l.$$

Given an orthonormal basis $v_{1m}, \ldots, v_{km}$ of $E_m^s(x)$ for each $m \in \mathbb{N}$, there exists a convergent subsequence, say with limits $v_1, \ldots, v_k$ that also form an orthonormal set. This shows that $G^s(x)$ contains a subspace $E^s(x)$ of dimension k (generated by $v_1, \ldots, v_k$). Analogously, one can show that $G^u(x)$ contains a subspace $E^u(x)$ of dimension l. On the other hand, it follows from (5.29) that

$$E^s(x) \cap E^u(x) \subset G^s(x) \cap G^u(x)$$
$$\subset C^s(x) \cap C^u(x) = \{0\}$$

since $\gamma < 1$. Moreover, it follows from (5.18) that

$$\dim M = \dim F^s(x) + \dim F^u(x)$$
$$= k + l$$
$$= \dim E^s(x) + \dim E^u(x)$$

and thus, the spaces $E^s(x)$ and $E^u(x)$ generate $T_x M$. Hence, we obtain the direct sum in (5.2).

Step 3. Estimates on the spaces $E^s(x)$ and $E^u(x)$.

Given $v \in E^s(x)$ and $n \in \mathbb{N}$, it follows from (5.21) together with (5.29) and (5.30) that

$$d_x f^k v \in d_x f^k G^s(x) = G^s\left(f^k(x)\right) \subset C^s\left(f^k(x)\right)$$

for $k = 0, \ldots, n$. Hence, it follows from (5.23) that

$$\|d_x f^n v\|' \leq \mu^n \|v\|'. \tag{5.32}$$

Analogously, given $v \in E^u(x)$ and $n \in \mathbb{N}$, it follows from (5.22) that

$$\|d_x f^{-n} v\|' \leq \mu^n \|v\|'. \tag{5.33}$$

Now we show that

$$E^s(x) = G^s(x) \quad \text{and} \quad E^u(x) = G^u(x)$$

for any $x \in \Lambda$. If there existed a $v \in G^s(x) \setminus E^s(x) \subset C^s(x)$, then one could write $v = v^s + v^u$, where $v^s \in E^s(x)$ and $v^u \in E^u(x) \setminus \{0\}$. For each $n \in \mathbb{N}$, we would have

$$\mu^{-n} \|v^u\|' \leq \|d_x f^n v^u\|'$$
$$\leq \|d_x f^n v\|' + \|d_x f^n v^s\|'$$
$$\leq \mu^n (\|v\|' + \|v^s\|').$$

This implies that

$$\|v^u\|' \leq \mu^{2n} (\|v\|' + \|v^s\|') \to 0$$

when $n \to \infty$ and thus $v^u = 0$. This contradiction shows that $E^s(x) = G^s(x)$. One can show in an analogous manner that $E^u(x) = G^u(x)$. Finally, it follows from (5.30) and (5.31) that

$$d_x f^{-1} E^s(x) = E^s(f^{-1}(x)) \quad \text{and} \quad d_x f E^u(x) = E^u(f(x)).$$

Therefore, Λ is a hyperbolic set, taking the constants $\lambda = \mu$ and $c = 1$, in view of (5.32) and (5.33). $\qquad \square$

5.4 Stability of Hyperbolic Sets

In this section we describe briefly some stability properties of a hyperbolic set under sufficiently small perturbations. In particular, we consider diffeomorphisms for which the whole manifold is a hyperbolic set.

Given differentiable maps $f, g \colon M \to M$, we define

$$d(f, g) = \sup_{x \in M} d(f(x), g(x)) + \sup_{x \in M} \|d_x f - d_x g\|. \tag{5.34}$$

Theorem 5.5 *Let Λ be a hyperbolic set for a C^1 diffeomorphism $f \colon M \to M$. Then there exist $\varepsilon > 0$ and an open set $U \supset \Lambda$ such that if $g \colon M \to M$ is a C^1 diffeomorphism with $d(f, g) < \varepsilon$ and $\Lambda' \subset U$ is a compact g-invariant set, then Λ' is a hyperbolic set for g.*

Proof By Theorem 5.1, the stable and unstable spaces $E^s(x)$ and $E^u(x)$ vary continuously with $x \in \Lambda$. Hence, by Tietze's extension theorem,[1] one can consider continuous extensions $F^s(x)$ and $F^u(x)$, respectively, of $E^s(x)$ and $E^u(x)$, for x in some open neighborhood U of Λ such that

$$T_x M = F^s(x) \oplus F^u(x) \quad \text{for } x \in U. \tag{5.35}$$

Given $\gamma > 0$, consider the cones $C^s(x)$ and $C^u(x)$ in (5.19) and (5.20) associated to the decompositions in (5.35). By Theorem 5.3, there exist constants $\mu, \gamma \in (0,1)$ and an inner product $\langle \cdot, \cdot \rangle' = \langle \cdot, \cdot \rangle'_x$ in $T_x M$ varying continuously with x such that:

1. for each $x \in \Lambda$,

$$d_x f \overline{C^u(x)} \subsetneq C^u(f(x)) \quad \text{and} \quad d_x f^{-1} \overline{C^s(x)} \subsetneq C^s\big(f^{-1}(x)\big);$$

2. for each $x \in \Lambda$,

$$\|d_x f v\|' > \mu^{-1} \|v\|' \quad \text{for } v \in \overline{C^u(x)} \setminus \{0\}$$

and

$$\left\| d_x f^{-1} v \right\|' > \mu^{-1} \|v\|' \quad \text{for } v \in \overline{C^s(x)} \setminus \{0\}.$$

Denoting by S_x the closed unit sphere in $T_x M$ (with respect to the norm $\|\cdot\| = \|\cdot\|'_x$), these properties are equivalent to:

1. for each $x \in \Lambda$,

$$d_x f\big(S_x \cap \overline{C^u(x)}\big) \subsetneq C^u(f(x)) \quad \text{and} \quad d_x f^{-1}\big(S_x \cap \overline{C^s(x)}\big) \subsetneq C^s\big(f^{-1}(x)\big);$$

2. for each $x \in \Lambda$,

$$\|d_x f v\|' > \mu^{-1} \quad \text{for } v \in S_x \cap \overline{C^u(x)}$$

and

$$\left\| d_x f^{-1} v \right\|' > \mu^{-1} \quad \text{for } v \in S_x \cap \overline{C^s(x)}.$$

Now we note that the set

$$\big\{(x,v) \in \Lambda \times T_x M : \|v\|'_x = 1 \big\}$$

is compact since the inner product $\langle \cdot, \cdot \rangle'_x$ and thus also the norm $\|\cdot\|'_x$ vary continuously with x. For any sufficiently small open neighborhood $U \supset \Lambda$, the properties above hold for any $x \in U$ (and some continuous extension of the inner product). Moreover, for any sufficiently small ε the same properties also hold for any $x \in U$

---

[1] **Theorem** (See for example [43]) If $f : A \to \mathbb{R}$ is a continuous function in a closed subset $A \subset X$ of a normal space (that is, a space such that any two disjoint closed sets have disjoint open neighborhoods), then there exists a continuous function $g : X \to \mathbb{R}$ such that $g|A = f$.

with f replaced by g. It follows from Theorem 5.4 that any compact g-invariant set $\Lambda' \subset U$ is a hyperbolic set for g. $\qquad\square$

Now we consider the particular case of the Anosov diffeomorphisms.

Definition 5.7 A diffeomorphism $f: M \to M$ of a compact manifold M is called an *Anosov diffeomorphism* if M is a hyperbolic set for f.

For example, any automorphism of the torus induced by a matrix without eigenvalues with modulus 1 (called a *hyperbolic* automorphism of the torus) is an Anosov diffeomorphism (see Exercise 5.15).

The following result is an immediate consequence of Theorem 5.5.

Theorem 5.6 *The set of Anosov diffeomorphisms of class* C^1 *of a compact manifold* M *is open with respect to the topology induced by the distance* d *in* (5.34).

5.5 Exercises

Exercise 5.1 Use Theorem 5.1 to show that if Λ is a hyperbolic set, then

$$\inf\{\angle(E^s(x), E^u(x)) : x \in \Lambda\} > 0.$$

Exercise 5.2 Find explicitly the 2-periodic points of the Smale horseshoe.

Exercise 5.3 Let $T: X \to X$ be a map of a complete metric space X. Show that if T^2 is a contraction, then T has a unique fixed point in X.

Exercise 5.4 Determine whether the set of all bounded C^1 functions $f: \mathbb{R} \to \mathbb{R}$ is a complete metric space with the distance

$$d(f, g) = \sup\{|f(x) - g(x)| : x \in \mathbb{R}\}. \tag{5.36}$$

Exercise 5.5 Determine whether the set of all bounded functions $f: \mathbb{R} \to \mathbb{R}$ such that

$$|f(x) - f(y)| \le |x - y| \quad \text{for } x, y \in \mathbb{R} \tag{5.37}$$

is a complete metric space with the distance d in (5.36).

Exercise 5.6 Repeat Exercise 5.5 with condition (5.37) replaced by

$$|f(x) - f(y)| \le c|x - y|^\alpha \quad \text{for } |x - y| \le d.$$

Exercise 5.7 Determine whether there exists a diffeomorphism $h: \mathbb{R} \to \mathbb{R}$ such that

$$h \circ f = g \circ h$$

for the maps:

1. $f(x) = 2x$ and $g(x) = 3x$;
2. $f(x) = 4x$ and $g(x) = -2x$;
3. $f(x) = 3x$ and $g(x) = x^3$.

Exercise 5.8 Give an example of a hyperbolic set for which not all spaces $E^s(x)$ have the same dimension.

Exercise 5.9 Determine whether S^1 is a hyperbolic set for some diffeomorphism of the circle.

Exercise 5.10 Show that the stable and unstable spaces of a hyperbolic set are uniquely determined.

Exercise 5.11 Find spaces $F^s(x)$, $F^u(x) \subset \mathbb{R}^2$ and an inner product $\langle \cdot, \cdot \rangle'_x$ such that

$$C^s(x) = \{0\} \cup \{(v, w) \in \mathbb{R}^2 : vw > 0\}.$$

Exercise 5.12 Show that if Λ is a hyperbolic set for a diffeomorphism f, then it is also a hyperbolic set for f^2.

Exercise 5.13 Identify the following statement as true or false: if Λ is a hyperbolic set for f^2, where f is a diffeomorphism, then Λ is a hyperbolic set for f.

Exercise 5.14 Let x be a fixed point of a diffeomorphism $f: M \to M$. Show that $\{x\}$ is a hyperbolic set for f if and only if the linear transformation

$$d_x f: T_x M \to T_x M$$

has no eigenvalues with modulus 1. Hint: Since $d_x f|T_x M$ can have nonreal eigenvalues, consider the complexification

$$T_x M^{\mathbb{C}} = \{u + iv : u, v \in T_x M\}$$

of the tangent space $T_x M$, with the norm

$$\|u + iv\| = \sqrt{\|u\|^2 + \|v\|^2} \quad \text{for } u, v \in T_x M,$$

and consider the extension $A: T_x M^{\mathbb{C}} \to T_x M^{\mathbb{C}}$ of $d_x f$ defined by

$$A(u + iv) = d_x f u + i d_x f v.$$

Exercise 5.15 Given an automorphism of the torus $T_A: \mathbb{T}^n \to \mathbb{T}^n$, show that $\mathbb{T}^n$ is a hyperbolic set for T_A if and only if A has no eigenvalues with modulus 1 (with the inner product in each tangent space $T_x \mathbb{T}^n$ induced from the standard inner product in $\mathbb{R}^n$). Hint: Consider the extension of $A \colon \mathbb{R}^n \to \mathbb{R}^n$ to $\mathbb{C}^n$ defined by

$$A(u + iv) = Au + iAv.$$

Exercise 5.16 Let x be a periodic point with period n of a diffeomorphism f. Give a necessary and sufficient condition in terms of $d_x f^n$ so that the periodic orbit

$$\{f^k(x) : k = 0, \ldots, n-1\}$$

is a hyperbolic set for f.

Exercise 5.17 Let $f \colon M \to M$ be a C^1 diffeomorphism and let $\Lambda \subset M$ be a compact f-invariant set. Show that Λ is a hyperbolic set for f if and only if there exist a decomposition (5.18) and an inner product $\langle \cdot, \cdot \rangle'_x$ in $T_x M$, for each $x \in \Lambda$, and a constant $\gamma \in (0, 1)$ such that:

1. for any $x \in \Lambda$,

$$d_x f C^u(x) \subset C^u(f(x)) \quad \text{and} \quad d_x f^{-1} C^s(x) \subset C^s(f^{-1}(x));$$

2. for any $x \in \Lambda$,

$$\|d_x f v\|' > \|v\|' \quad \text{for } v \in \overline{C^u(x)} \setminus \{0\}$$

and

$$\|d_x f^{-1} v\|' > \|v\|' \quad \text{for } v \in \overline{C^s(x)} \setminus \{0\}.$$

Exercise 5.18 Consider the set

$$N = S^1 \times \{(x, y) \in \mathbb{R}^2 : x^2 + y^2 \le 1\}$$

and the map $f \colon N \to N$ defined by

$$f(\theta, x, y) = \left(2\theta, \lambda x + \frac{1}{2} \cos(2\pi\theta), \mu y + \frac{1}{2} \sin(2\pi\theta)\right)$$

for some constants $\lambda, \mu \in (0, 1/2)$. Show that:

1. the map f is one-to-one;
2. the *solenoid* $\Lambda = \bigcap_{n \in \mathbb{N}} f^n(N)$ is a compact f-invariant set;
3. the restriction $f|U \colon U \to f(U)$ of f to the set

$$U = S^1 \times \{(x, y) \in \mathbb{R}^2 : x^2 + y^2 < 1\}$$

   is a diffeomorphism;
4. Λ is a hyperbolic set for f.

Exercise 5.19 Determine whether the map $f|\Lambda$ in Exercise 5.18 has periodic points with period 3.

Exercise 5.20 Show that the periodic points of the Smale horseshoe Λ are dense in Λ.

Chapter 6
Hyperbolic Dynamics II

This chapter is a natural continuation of the former chapter. We consider topics that, in view of the necessary details or in view of the need for results from other areas, may be considered less elementary. We first describe the behavior of the orbits of a diffeomorphism near a hyperbolic fixed point. More precisely, we establish two fundamental results of hyperbolic dynamics: the Grobman–Hartman theorem and the Hadamard–Perron theorem. We also establish the existence of stable and unstable manifolds for all points of a hyperbolic set and we show how they give rise to a local product structure for any locally maximal hyperbolic set. We conclude the chapter with an introduction to geodesic flows on surfaces of constant negative curvature and their hyperbolicity. In particular, we consider isometries, Möbius transformations, geodesics, quotients by isometries and the construction of compact surfaces of genus at least 2.

6.1 Behavior Near a Hyperbolic Fixed Point

Let x be a fixed point of a diffeomorphism f. When $\{x\}$ is a hyperbolic set for f, we say that x is a *hyperbolic fixed point*. In this section we describe the behavior of the orbits of a diffeomorphism in an open neighborhood of a hyperbolic fixed point. For simplicity of exposition, we consider only diffeomorphisms on $\mathbb{R}^p$.

6.1.1 The Grobman–Hartman Theorem

We first establish a result showing that in a sufficiently small open neighborhood of a hyperbolic fixed point x the orbits of f are obtained from the orbits of the linear transformation $d_x f$ applying a homeomorphism. This corresponds to the notion of a local topological conjugacy (compare with Definition 3.9).

L. Barreira, C. Valls, *Dynamical Systems*, Universitext, DOI 10.1007/978-1-4471-4835-7_6, 113
© Springer-Verlag London 2013

Definition 6.1 Two maps $f\colon X \to X$ and $g\colon Y \to Y$, where X and Y are topological spaces, are said to be *(locally) topologically conjugate*, respectively, in open sets $U \subset X$ and $V \subset Y$ if there exists a homeomorphism $h\colon U \to V$ with $h(U) = V$ such that $h \circ f = g \circ h$ in U.

The Grobman–Hartman theorem establishes the existence of a (local) topological conjugacy between f and $d_x f$ in open neighborhoods, respectively, of x and 0.

Theorem 6.1 (Grobman–Hartman) *Let $x \in \mathbb{R}^p$ be a hyperbolic fixed point of a C^1 diffeomorphism $f\colon \mathbb{R}^p \to \mathbb{R}^p$. Then there exists a homeomorphism $h\colon U \to V$ with $h(U) = V$, where U and V are, respectively, open neighborhoods of x and 0, such that*

$$h \circ f = d_x f \circ h \quad in\ U. \tag{6.1}$$

Proof We divide the proof into steps.

Step 1. Preliminaries.

Without loss of generality, one can always assume that $x = 0$. Indeed, the map $\bar{f}\colon \mathbb{R}^p \to \mathbb{R}^p$ defined by

$$\bar{f}(y) = f(y + x) - f(x)$$

is also a diffeomorphism and it satisfies

$$\bar{f}(0) = 0 \quad \text{and} \quad d_0 \bar{f} = d_x f.$$

In particular, 0 is a hyperbolic fixed point of $\bar{f}$. Now we modify the diffeomorphism f outside an open neighborhood of 0 (already assuming that 0 is a hyperbolic fixed point of f). More precisely, given $\delta > 0$, take $r \in (0, 1)$ so small that

$$\sup\{\|d_y f - d_0 f\| : y \in B(0, r)\} \le \delta/3 \tag{6.2}$$

(recall that the function $y \mapsto d_y f$ is continuous). Consider also a C^1 function $\alpha\colon \mathbb{R}^p \to [0, 1]$ such that:

1. $\alpha(y) = 1$ for $y \in B(0, r/3)$;
2. $\alpha(y) = 0$ for $y \in \mathbb{R}^p \setminus B(0, r)$;
3. $\sup\{\|d_y \alpha\| : y \in \mathbb{R}^p\} \le 2/r$.

We define a map $g\colon \mathbb{R}^p \to \mathbb{R}^p$ by

$$g(y) = Ay + \alpha(y)F(y),$$

where

$$A = d_0 f \quad \text{and} \quad F(y) = f(y) - Ay.$$

Note that

$$g(y) = Ay + F(y) = f(y)$$

for $y \in B(0, r/3)$, that is, g coincides with f on the ball $B(0, r/3)$. Since $F(0) = 0$, it follows from (6.2) and the mean value theorem that

$$\sup\{\|F(y)\| : y \in B(0, r)\} \le \delta r/3.$$

Since $r < 1$ and the function α is zero outside the ball $B(0, r)$, we have

$$\sup\{\|\alpha(y)F(y)\| : y \in B(0, r)\} \le \delta r/3 < \delta. \tag{6.3}$$

Moreover,

$$\|d_y(\alpha F)\| = \|d_y \alpha F(y) + \alpha(y) d_y F\|$$
$$\le \sup_{y \in \mathbb{R}^p} \|d_y \alpha\| \sup_{y \in B(0,r)} \|F(y)\| + \sup_{y \in \mathbb{R}^p} \|d_y F\|$$
$$< \frac{2}{r} \cdot \frac{\delta r}{3} + \frac{\delta}{3} = \delta$$

and it follows again from the mean value theorem that

$$\|\alpha(y)F(y) - \alpha(z)F(z)\| \le \delta \|y - z\| \tag{6.4}$$

for $y, z \in \mathbb{R}^p$.

Step 2. Construction of a norm.

Now we consider a norm that is analogous to the one introduced in the proof of Theorem 5.3. Namely, given $m \in \mathbb{N}$ such that $c\lambda^m < 1$, for each $v \in \mathbb{R}^p$, let

$$\|v\|' = \max\{\|v^s\|', \|v^u\|'\},$$

where $v = v^s + v^u$, with $v^s \in E^s(0)$ and $v^u \in E^u(0)$, and where

$$(\|v^s\|')^2 = \sum_{n=0}^{m-1} \|A^n v^s\|^2 \tag{6.5}$$

and

$$(\|v^u\|')^2 = \sum_{n=0}^{m-1} \|A^{-n} v^u\|^2. \tag{6.6}$$

It follows from (6.5) and (6.6) that

$$\|v^s\| \le \|v^s\|' \le C\|v^s\| \tag{6.7}$$

and

$$\|v^u\| \le \|v^u\|' \le C\|v^u\|, \tag{6.8}$$

where $C = c\sqrt{m}$ (see (5.25)). Moreover, it follows from (5.26) and (5.27) that

$$\|Av^s\|' \le \tau\|v^s\|' \quad \text{and} \quad \|A^{-1}v^u\|' \le \tau\|v^u\|', \tag{6.9}$$

where

$$\tau = \sqrt{1 - \frac{1 - c^2\lambda^{2m}}{c^2 m}} < 1.$$

Step 3. Formulation of an abstract problem.

Let X be the space of bounded continuous functions $v \colon \mathbb{R}^p \to \mathbb{R}^p$ with $v(0) = 0$. One can easily verify that X is a Banach space (that is, a complete normed space) with the norm

$$\|v\|_\infty = \max\{\|v_s\|_\infty, \|v_u\|_\infty\},$$

where

$$v(y) = \bigl(v_s(y), v_u(y)\bigr) \in E^s(0) \oplus E^u(0)$$

and

$$\|v_s\|_\infty = \sup\{\|v_s(y)\|', \ y \in \mathbb{R}^p\}, \qquad \|v_u\|_\infty = \sup\{\|v_u(y)\|', \ y \in \mathbb{R}^p\}.$$

We write

$$A_s = A|E^s(0), \qquad A_u = A|E^u(0), \tag{6.10}$$

and we consider functions $G, H \colon \mathbb{R}^p \to \mathbb{R}^p$ with $G(0) = H(0) = 0$ such that

$$\|G(y)\| \le \delta, \qquad \|H(y)\| \le \delta \tag{6.11}$$

and

$$\|G(y) - G(z)\| \le \delta\|y - z\|, \qquad \|H(y) - H(z)\| \le \delta\|y - z\|, \tag{6.12}$$

for $y, z \in \mathbb{R}^p$. We note that, by (6.3) and (6.4), the function $G = \alpha F$ satisfies these properties. Writing

$$G(y) = \bigl(G_s(y), G_u(y)\bigr) \in E^s(0) \oplus E^u(0)$$

and

$$H(y) = \bigl(H_s(y), H_u(y)\bigr) \in E^s(0) \oplus E^u(0),$$

it follows from (6.11) together with (6.7) and (6.8) that

$$\|G_s(y)\|' \le C\|G_s(y)\| \le C\|G(y)\| \le C\delta \tag{6.13}$$

and analogously,

$$\|G_u(y)\|' \leq C\|G_u(y)\| \leq C\|G(y)\| \leq C\delta, \tag{6.14}$$

with identical inequalities for H_s and H_u. On the other hand, since

$$\sqrt{a^2 + b^2} \leq \sqrt{2 \max\{a^2, b^2\}} \leq \sqrt{2} \max\{|a|, |b|\},$$

it follows from (6.12) together with (6.7) and (6.8) that

$$\|G_s(y) - G_s(z)\|' \leq C\|G_s(y) - G_s(z)\| \leq C\|G(y) - G(z)\|$$
$$\leq C\delta\|y - z\| \leq C\delta\sqrt{2} \max\{\|y^s - z^s\|, \|y^u - z^u\|\}$$
$$\leq C\delta\sqrt{2} \max\{\|y^s - z^s\|', \|y^u - z^u\|'\} = C\delta\sqrt{2}\|y - z\|' \tag{6.15}$$

and analogously,

$$\|G_u(y) - G_u(z)\|' \leq C\delta\sqrt{2}\|y - z\|', \tag{6.16}$$

again with identical inequalities for H_s and H_u.

Now we consider the equation

$$(A + G) \circ h = h \circ (A + H), \tag{6.17}$$

where $h = \mathrm{Id} + v$. It is equivalent to the system of equations

$$A_s \circ h_s + G_s \circ h = h_s \circ (A + H),$$
$$A_u \circ h_u + G_u \circ h = h_u \circ (A + H). \tag{6.18}$$

Since

$$A_s \circ h_s = A_s + A_s \circ v_s \quad \text{and} \quad h_s \circ (A + H) = A_s + H_s + v_s \circ (A + H),$$

the first equation in (6.18) is equivalent to

$$A_s \circ v_s + G_s \circ h = H_s + v_s \circ (A + H). \tag{6.19}$$

Analogously, the second equation in (6.18) is equivalent to

$$A_u \circ v_u + G_u \circ h = H_u + v_u \circ (A + H). \tag{6.20}$$

Now we show that the functions $A + G$ and $A + H$ are invertible for any sufficiently small δ. It follows from (6.4) that

$$\|(A + G)(y) - (A + G)(z)\| = \|A(y - z) + G(y) - G(z)\|$$
$$\geq \|A^{-1}\|^{-1} \cdot \|y - z\| - \|G(y) - G(z)\|$$
$$\geq (\|A^{-1}\|^{-1} - \delta)\|y - z\|$$

and thus, the function $A + G$ is invertible for $\delta < \|A^{-1}\|^{-1}$. Likewise for the function $A + H$. This allows us to write Eqs. (6.19) and (6.20) in the form

$$
\begin{aligned}
v_s &= \big(A_s \circ v_s + G_s \circ (\mathrm{Id} + v) - H_s\big) \circ (A + H)^{-1}, \\
v_u &= A_u^{-1} \circ \big(v_u \circ (A + H) + H_u - G_u \circ (\mathrm{Id} + v)\big).
\end{aligned}
\tag{6.21}
$$

Step 4. Existence of a fixed point.

We define a map T in the space X by $T(v) = w = (w_s, w_u)$, where

$$
\begin{aligned}
w_s &= \big(A_s \circ v_s + G_s \circ (\mathrm{Id} + v) - H_s\big) \circ (A + H)^{-1}, \\
w_u &= A_u^{-1} \circ \big(v_u \circ (A + H) + H_u - G_u \circ (\mathrm{Id} + v)\big).
\end{aligned}
$$

We note that v is a fixed point of T, that is, $T(v) = v$ if and only if system (6.21) holds, which is equivalent to identity (6.17).

Now we show that $T(X) \subset X$ and that T is a contraction. It follows from (6.9), (6.13) and (6.14), together with the analogous inequalities for H_s and H_u, that

$$
\|w_s\|_\infty \le \tau \|v_s\|_\infty + 2C\delta < +\infty
$$

and

$$
\|w_u\|_\infty \le \tau \|v_u\|_\infty + 2\tau C\delta < +\infty.
$$

Moreover, $w(0) = 0$ and $w = (w_s, w_u) \in X$. On the other hand, given functions $v = (v_s, v_u), \bar{v} = (\bar{v}_s, \bar{v}_u) \in X$, we have

$$
\begin{aligned}
T(v) - T(\bar{v}) = \big(&[A_s \circ (v_s - \bar{v}_s) + G_s \circ (\mathrm{Id} + v) - G_s \circ (\mathrm{Id} + \bar{v})] \circ (A + H)^{-1}, \\
&A_u^{-1} \circ [(v_u - \bar{v}_u) \circ (A + H) - G_u \circ (\mathrm{Id} + v) + G_u \circ (\mathrm{Id} + \bar{v})]\big).
\end{aligned}
$$

It follows from (6.9) and (6.10) that

$$
\big\|\big(A_s \circ (v_s - \bar{v}_s) \circ (A + H)^{-1}\big)(y)\big\|' \le \|A_s \circ (v_s - \bar{v}_s)\|_\infty \le \tau \|v_s - \bar{v}_s\|_\infty
$$

and

$$
\big\|\big(A_u^{-1} \circ (v_u - \bar{v}_u) \circ (A + H)\big)(y)\big\|' \le \big\|A_u^{-1} \circ (v_u - \bar{v}_u)\big\|_\infty \le \tau \|v_u - \bar{v}_u\|_\infty.
$$

Moreover, it follows from (6.15) and (6.16) that

$$
\big\|G_s \circ (\mathrm{Id} + v) - G_s \circ (\mathrm{Id} + \bar{v})\big\|_\infty \le C\delta\sqrt{2}\|v - \bar{v}\|_\infty
$$

and

$$
\big\|-G_u \circ (\mathrm{Id} + v) + G_u \circ (\mathrm{Id} + \bar{v})\big\|_\infty \le C\delta\sqrt{2}\|v - \bar{v}\|_\infty.
$$

Hence, we obtain

$$\|T(v) - T(\bar{v})\|_\infty$$
$$\leq \max\{\tau\|v_s - \bar{v}_s\|_\infty + C\delta\sqrt{2}\|v - \bar{v}\|_\infty, \tau\|v_u - \bar{v}_u\|_\infty + \tau C\delta\sqrt{2}\|v - \bar{v}\|_\infty\}$$
$$\leq (\tau + C\delta\sqrt{2})\|v - \bar{v}\|_\infty.$$

If necessary, one can go back and take $\delta > 0$ such that $\tau + C\delta\sqrt{2} < 1$, which guarantees that T is a contraction in the Banach space X. By the contraction mapping principle,[1] there exists a unique function $v \in X$ such that $T(v) = v$, that is, such that Eq. (6.17) holds with $h = \text{Id} + v$.

Step 5. Conclusion of the proof.

Finally, we consider several particular cases of Eq. (6.17). Taking $G = 0$ and $H = \alpha F$ (which is possible in view of (6.3) and (6.4)), we obtain a unique function $v \in X$ such that $h = \text{Id} + v$ satisfies

$$A \circ h = h \circ g \tag{6.22}$$

(we recall that $g = A + \alpha F$). Similarly, taking $G = \alpha F$ and $H = 0$, we obtain a unique function $\bar{v} \in X$ such that $\bar{h} = \text{Id} + \bar{v}$ satisfies

$$g \circ \bar{h} = \bar{h} \circ A. \tag{6.23}$$

We show that

$$h \circ \bar{h} = \bar{h} \circ h = \text{Id}.$$

It follows from (6.22) and (6.23) that

$$A \circ (h \circ \bar{h}) = h \circ g \circ \bar{h} = (h \circ \bar{h}) \circ A \tag{6.24}$$

and

$$(\bar{h} \circ h) \circ g = \bar{h} \circ A \circ h = g \circ (\bar{h} \circ h). \tag{6.25}$$

Since

$$\|h \circ \bar{h} - \text{Id}\|_\infty = \left\|(\text{Id} + v) \circ (\text{Id} + \bar{v}) - \text{Id}\right\|_\infty$$
$$= \left\|\bar{v} + v \circ (\text{Id} + \bar{v})\right\|_\infty$$
$$\leq \|\bar{v}\|_\infty + \|v\|_\infty < +\infty$$

and

$$\|\bar{h} \circ h - \text{Id}\|_\infty \leq \|v\|_\infty + \|\bar{v}\|_\infty < +\infty,$$

---

[1]**Theorem** (See for example [12]) If $T: X \to X$ is a contraction (that is, there exists a $\lambda \in (0, 1)$ such that $d(T(x), T(y)) \leq \lambda d(x, y)$ for $x, y \in X$) in a complete metric space, then T has a unique fixed point.

the continuous functions $h \circ \bar{h} - \mathrm{Id}$ and $\bar{h} \circ h - \mathrm{Id}$ belong to X. It follows from (6.24) and (6.25) together with the uniqueness of the solutions $v \in X$ of the equations

$$A \circ (\mathrm{Id} + v) = (\mathrm{Id} + v) \circ A \quad \text{and} \quad g \circ (\mathrm{Id} + v) = (\mathrm{Id} + v) \circ g, \qquad (6.26)$$

taking, respectively, $G = H = 0$ and $G = H = \alpha F$, that

$$h \circ \bar{h} - \mathrm{Id} = \bar{h} \circ h - \mathrm{Id} = 0.$$

Indeed, $v = 0$ is a solution of both equations in (6.26). This shows that h is a homeomorphism, with inverse $\bar{h}$. Since $g = f$ in $B(0, r/3)$, it follows from (6.22) that property (6.1) holds for $U = B(0, r/3)$. $\square$

Theorem 6.1 establishes a precise relation between the orbits of f and the orbits of the linear transformation $A = d_x f$. Namely, it follows from (6.1) that

$$h\big(f^n(y)\big) = A^n(h(y)) \qquad (6.27)$$

whenever

$$n \in \mathbb{Z}^+ \quad \text{and} \quad y \in \bigcap_{m=0}^{n-1} f^{-m} U$$

or

$$n \in \mathbb{Z}^- \quad \text{and} \quad y \in \bigcap_{m=0}^{n-1} f^m U.$$

In other words, the points of the orbit $\gamma_f(y)$ that are in U are mapped by the homeomorphism h to corresponding points of the orbit $\gamma_A(h(y))$, that is,

$$h\big(\gamma_f(y) \cap U\big) = \gamma_A(h(y)) \cap V. \qquad (6.28)$$

We also have

$$h\big(\gamma_f^+(y) \cap U\big) = \gamma_A^+(h(y)) \cap V \quad \text{and} \quad h\big(\gamma_f^-(y) \cap U\big) = \gamma_A^-(h(y)) \cap V$$

for each $y \in \mathbb{R}^p$.

6.1.2 The Hadamard–Perron Theorem

In this section we continue the study of the behavior of the orbits of a diffeomorphism in an open neighborhood of a hyperbolic fixed point.

Let $f : \mathbb{R}^p \to \mathbb{R}^p$ be a C^1 diffeomorphism and let $x \in \mathbb{R}^p$ be a hyperbolic fixed point of f. It follows from the Grobman–Hartman theorem (Theorem 6.1) that there exists a homeomorphism $h: U \to V$ with $h(U) = V$, where U and V are, respectively, open neighborhoods of x and 0, such that

$$h \circ f = A \circ h \quad \text{in } U, \tag{6.29}$$

where $A = d_x f$. In view of (6.28), identity (6.29) establishes a relation between the orbits of f and the orbits of A.

Now let $E^s(x)$ and $E^u(x)$ be the stable and unstable spaces at x. We have

$$E^s(x) = \left\{ y \in \mathbb{R}^p : A^n y \to 0 \text{ when } n \to +\infty \right\} \tag{6.30}$$

and

$$E^u(x) = \left\{ y \in \mathbb{R}^p : A^n y \to 0 \text{ when } n \to -\infty \right\}. \tag{6.31}$$

Moreover, one can always assume that the neighborhood V is such that

$$A\left(E^s(x) \cap V\right) \subset E^s(x) \cap V \tag{6.32}$$

and

$$A^{-1}\left(E^u(x) \cap V\right) \subset E^u(x) \cap V. \tag{6.33}$$

Indeed, let $\langle \cdot, \cdot \rangle'$ be the inner product constructed in the proof of Theorem 5.3 with $d_x f$ replaced by A (and thus with $d_x f^n$ replaced by A^n). By (5.26) and (5.27), given $r > 0$, any open neighborhood V of 0 such that

$$E^s(x) \cap V = \left\{ v \in E^s(x) : \|v\|' < r \right\}$$

and

$$E^u(x) \cap V = \left\{ v \in E^u(x) : \|v\|' < r \right\}$$

satisfies (6.32) and (6.33). It follows from (6.30) and (6.31) together with (6.32) and (6.33) (using, for example, the Jordan form of A) that

$$E^s(x) \cap V = \left\{ y \in V : A^n y \in V \text{ for } n > 0 \right\} \tag{6.34}$$

and

$$E^u(x) \cap V = \left\{ y \in V : A^n y \in V \text{ for } n < 0 \right\}. \tag{6.35}$$

We also consider the sets

$$V^s(x) = h^{-1}\left(E^s(x) \cap V\right) \quad \text{and} \quad V^u(x) = h^{-1}\left(E^u(x) \cap V\right)$$

(both are contained in U). We note that $x \in V^s(x) \cap V^u(x)$. It follows from (6.34) and (6.35) together with (6.27) that

$$V^s(x) = \left\{ y \in U : f^n(y) \in U \text{ for } n > 0 \right\}$$

and

$$V^u(x) = \{y \in U : f^n(y) \in U \text{ for } n < 0\}.$$

Combining these identities with (6.32) and (6.33), we conclude that

$$f(V^s(x)) \subset V^s(x) \quad \text{and} \quad f^{-1}(V^u(x)) \subset V^u(x). \qquad (6.36)$$

The Hadamard–Perron theorem says that the sets $V^s(x)$ and $V^u(x)$ are manifolds, tangent, respectively, to the spaces $E^s(x)$ and $E^u(x)$.

Theorem 6.2 (Hadamard–Perron) *Let $x \in \mathbb{R}^p$ be a hyperbolic fixed point of a C^1 diffeomorphism $f \colon \mathbb{R}^p \to \mathbb{R}^p$. Then there exists an open neighborhood B of x such that the sets $V^s(x) \cap B$ and $V^u(x) \cap B$ are manifolds of class C^1 with*

$$T_x(V^s(x) \cap B) = E^s(x) \quad \text{and} \quad T_x(V^u(x) \cap B) = E^u(x). \qquad (6.37)$$

For simplicity of exposition, we divide the proof into two steps.

Proposition 6.1 *Let $x \in \mathbb{R}^p$ be a hyperbolic fixed point of a C^1 diffeomorphism f. Then there exists an open neighborhood B of x such that the sets $V^s(x) \cap B$ and $V^u(x) \cap B$ are graphs of Lipschitz functions.*

Proof We establish the result only for $V^s(x)$ since the argument for $V^u(x)$ is entirely analogous. Again we divide the proof into steps.

Step 1. Preliminaries.

Consider the map $F \colon \mathbb{R}^p \to \mathbb{R}^p$ defined by

$$F(y) = f(y + x) - f(x).$$

It can be written in the form

$$F(v, w) = (Av + g(v, w), Bw + h(v, w)) \in E^s(x) \oplus E^u(x) \qquad (6.38)$$

for each $(v, w) \in E^s(x) \oplus E^u(x)$, where

$$A = d_x f | E^s(x) \colon E^s(x) \to E^s(x)$$

and

$$B = d_x f | E^u(x) \colon E^u(x) \to E^u(x).$$

We note that A and B are invertible linear transformations and that g and h are C^1 functions with

$$g(0) = 0, \qquad h(0) = 0, \qquad d_0 g = 0, \qquad d_0 h = 0.$$

Proceeding as in Step 1 of the proof of Theorem 5.3, one can always assume that the inner product $\langle \cdot, \cdot \rangle_x$ on the tangent space $T_x \mathbb{R}^p$ is such that

$$\|A\| < \tau \quad \text{and} \quad \|B^{-1}\| < \tau$$

for some constant $\tau \in (0, 1)$ (independent of x). Moreover,

$$(g, h)(y) = f(y + x) - f(x) - d_x f y$$

and the function

$$G \colon y \mapsto d_y(g, h) = d_{y+x} f - d_x f$$

is continuous since f is of class C^1. Given $\varepsilon > 0$, let

$$D_\varepsilon(x) = \overline{B_{2\varepsilon}^s(x)} \oplus \overline{B_{2\varepsilon}^u(x)},$$

where

$$B_{2\varepsilon}^s(x) \subset E^s(x) \quad \text{and} \quad B_{2\varepsilon}^u(x) \subset E^u(x)$$

are the open balls of radius 2ε centered at the origin. Since $G(0) = 0$, we have

$$K := \sup\{\|G(y)\| : y \in D_\varepsilon(x)\} \to 0$$

when $\varepsilon \to 0$. We also have

$$\|d_{(v,w)}g\| \le K \quad \text{and} \quad \|d_{(v,w)}h\| \le K \tag{6.39}$$

for $(v, w) \in \overline{D_\varepsilon^s} \oplus \overline{D_\varepsilon^u}$, where

$$D_\varepsilon^s = B_\varepsilon^s(x) \quad \text{and} \quad D_\varepsilon^u = B_\varepsilon^u(x). \tag{6.40}$$

Step 2. Formulation of an abstract problem.

Given $\sigma \in (0, 1]$, let X be the space of functions $\varphi \colon D_\varepsilon^s \to E^u(x)$ such that $\varphi(0) = 0$ and

$$\|\varphi(v) - \varphi(w)\| \le \sigma \|v - w\| \tag{6.41}$$

for $v, w \in D_\varepsilon^s$. One can easily verify that X is a complete metric space with the distance

$$d(\varphi, \psi) = \sup\{\|\varphi(v) - \psi(v)\| : v \in D_\varepsilon^s\}.$$

We want to show that there exists a function $\varphi \in X$ such that

$$V^s(x) \supset x + \{(v, \varphi(v)) : v \in D_\varepsilon^s\}. \tag{6.42}$$

We first proceed formally, assuming that (6.42) holds. It follows from (6.38) and the invariance property in (6.36) that

$$F\big(v, \varphi(v)\big) = \big(Av + g\big(v, \varphi(v)\big), B\varphi(v) + h\big(v, \varphi(v)\big)\big)$$

and

$$B\varphi(v) + h\big(v, \varphi(v)\big) = \varphi\big(Av + g\big(v, \varphi(v)\big)\big).$$

Thus,

$$\varphi(v) = B^{-1}\varphi\big(Av + g\big(v, \varphi(v)\big)\big) - B^{-1}h\big(v, \varphi(v)\big).$$

This leads us to introduce a map T on X by

$$T(\varphi)(v) = B^{-1}\varphi\big(Av + g\big(v, \varphi(v)\big)\big) - B^{-1}h\big(v, \varphi(v)\big).$$

We note that φ is a fixed point of T if and only if the graph of the function φ in (6.42) satisfies the invariance property in (6.36). Hence, if we show that T has a unique fixed point $\varphi \in X$, then the set on the right-hand side of (6.42) coincides with $V^s(x)$ in some open neighborhood of x, which yields the desired result.

Step 3. Existence of a fixed point.

We first show that the map T is well defined. This amounts to verifying that each point $Av + g(v, \varphi(v))$ is in the domain of φ, that is,

$$\big\|Av + g\big(v, \varphi(v)\big)\big\| < \varepsilon.$$

We have

$$\|\varphi(v)\| = \|\varphi(v) - \varphi(0)\| \le \sigma \|v\| \le \varepsilon$$

and it follows from (6.39) that

$$\big\|Av + g\big(v, \varphi(v)\big)\big\| \le \tau \|v\| + K \big\|(v, \varphi(v))\big\|$$
$$\le \big[\tau + K(1 + \sigma)\big]\|v\|$$
$$\le \big[\tau + K(1 + \sigma)\big]\varepsilon.$$

Since $K \to 0$ when $\varepsilon \to 0$, taking ε sufficiently small, we obtain

$$\big[\tau + K(1 + \sigma)\big]\varepsilon < \varepsilon$$

and hence, the map T is well defined.

Now we show that $T(X) \subset X$. We first observe that $T(\varphi)(0) = 0$ since $g(0) = 0$ and $h(0) = 0$. Moreover, given $v, w \in D_\varepsilon^s$, we have

$$
\begin{aligned}
\big\| T(\varphi)(v) - T(\varphi)(w) \big\| &\leq \tau \big\| \varphi\big(Av + g(v, \varphi(v))\big) - \varphi\big(Aw + g(w, \varphi(w))\big) \big\| \\
&\quad + \tau \big\| h(v, \varphi(v)) - h(w, \varphi(w)) \big\| \\
&\leq \tau\sigma \big\| A(v - w) + g(v, \varphi(v)) - g(w, \varphi(w)) \big\| \\
&\quad + \tau K \big\| (v, \varphi(v)) - (w, \varphi(w)) \big\| \\
&\leq \tau^2 \sigma \| v - w \| + \tau(1 + \sigma) K \big\| (v, \varphi(v)) - (w, \varphi(w)) \big\| \\
&\leq \big[\tau^2 \sigma + \tau(1 + \sigma)^2 K \big] \| v - w \|.
\end{aligned}
$$

Since $K \to 0$ when $\varepsilon \to 0$, taking ε sufficiently small, we obtain

$$
\tau^2 \sigma + \tau(1 + \sigma)^2 K < \sigma
$$

and hence,

$$
\big\| T(\varphi)(v) - T(\varphi)(w) \big\| \leq \sigma \| v - w \|
$$

for $v, w \in D_\varepsilon^s$. This shows that $T(X) \subset X$.

Finally, we show that T is a contraction. Given $\varphi, \psi \in X$ and $v \in D_\varepsilon^s$, we have

$$
\begin{aligned}
\big\| T(\varphi)(v) - T(\psi)(v) \big\| &\leq \tau \big\| \varphi\big(Av + g(v, \varphi(v))\big) - \psi\big(Av + g(v, \psi(v))\big) \big\| \\
&\quad + \tau \big\| h(v, \varphi(v)) - h(v, \psi(v)) \big\| \\
&\leq \tau \big\| \varphi\big(Av + g(v, \varphi(v))\big) - \varphi\big(Av + g(v, \psi(v))\big) \big\| \\
&\quad + \tau \big\| \varphi\big(Av + g(v, \psi(v))\big) - \psi\big(Av + g(v, \psi(v))\big) \big\| \\
&\quad + \tau K \big\| \varphi(v) - \psi(v) \big\| \\
&\leq \tau\sigma \big\| g(v, \varphi(v)) - g(v, \psi(v)) \big\| + \tau d(\varphi, \psi) + \tau K d(\varphi, \psi) \\
&\leq \big[\tau + \tau(1 + \sigma) K \big] d(\varphi, \psi).
\end{aligned}
$$

Taking ε sufficiently small so that

$$
\tau + \tau(1 + \sigma) K < 1,
$$

the map T is a contraction in the complete metric space X. Thus, T has a unique fixed point $\varphi \in X$. $\qquad\square$

Now we show that the function φ obtained in the proof of Proposition 6.1 is of class C^1.

Proposition 6.2 *Let $x \in \mathbb{R}^p$ be a hyperbolic fixed point of a C^1 diffeomorphism f. Then there exists an open neighborhood B of x such that the sets $V^s(x) \cap B$ and $V^u(x) \cap B$ are manifolds of class C^1 and satisfy (6.37).*

Fig. 6.1 The vector
$u = (w, \varphi(w)) - (v, \varphi(v))$

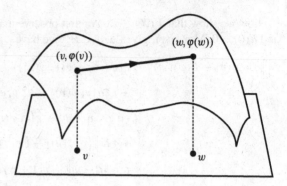

Proof We divide the proof into steps.

Step 1. Preliminaries.

Given $v, w \in D_\varepsilon^s$ with $v \neq w$, let

$$\Delta_{v,w} = \frac{(w, \varphi(w)) - (v, \varphi(v))}{\|(w, \varphi(w)) - (v, \varphi(v))\|}.$$

(see Fig. 6.1, where $\Delta_{v,w} = u/\|u\|$). We denote by S_v the set of all vectors $w \in T_x M$ such that $\Delta_{v,v_m} \to w$ when $m \to \infty$, for some sequence $(v_m)_{m \in \mathbb{N}}$ converging to v. When φ is differentiable at v, each of these vectors w is tangent to the graph of φ at the point $(v, \varphi(v))$ and has norm 1. We also consider the set

$$\tau_{(v,\varphi(v))} V^s = \{ \lambda w : w \in S_v, \lambda \in \mathbb{R} \}, \tag{6.43}$$

where

$$V^s = \{ (v, \varphi(v)) : v \in D_\varepsilon^s \}.$$

Lemma 6.1 *The function φ is differentiable at v if and only if $\tau_{(v,\varphi(v))} V^s$ is a subspace of dimension $\dim E^s(x)$.*

Proof By the former discussion, if φ is differentiable at v, then the vectors λw in (6.43) are exactly the elements of the tangent space to the graph of φ at the point $(v, \varphi(v))$. Hence, $\tau_{(v,\varphi(v))} V^s$ coincides with this tangent space and has dimension $\dim E^s(x)$.

On the other hand, if $\tau_{(v,\varphi(v))} V^s$ is a subspace of dimension $\dim E^s(x)$, then for each vector $u \in D_\varepsilon^s \setminus \{0\}$ with $v + u \in D_\varepsilon^s$, the limit

$$C(v, u) := \lim_{m \to \infty} \Delta_{v, v+s_m u} \tag{6.44}$$

exists for any sequence s_m such that $s_m \to 0$ when $m \to \infty$. Moreover, $C(u, v)$ is independent of the sequence s_m. Indeed, since each vector $\Delta_{v, v+s_m u}$ has norm 1, it follows from the compactness of the closed unit sphere of $\mathbb{R}^p$ that the sequence $(\Delta_{v, v+s_m u})_m$ has sublimits. Moreover, since $\tau_{(v,\varphi(v))} V^s$ is a subspace, for each u

there exists a unique vector $w = w_{v,u} \in E^u(x)$ such that $(u, w) \in \tau_{(v,\varphi(v))} V^s$. This implies that the limit in (6.44) exists and

$$C(v, u) = (u, w_{v,u}) / \|(u, w_{v,u})\|.$$

In other words, $d_v \varphi u = w_{v,u}$ and the function φ is differentiable at v. $\qquad \square$

Now we observe that $\Delta_{v,v_m} \to w$ when $m \to \infty$ if and only if

$$\lim_{m \to \infty} \frac{F(v_m, \varphi(v_m)) - F(v, \varphi(v))}{\|F(v_m, \varphi(v_m)) - F(v, \varphi(v))\|} = \frac{d_{(v,\varphi(v))} F w}{\|d_{(v,\varphi(v))} F w\|}.$$

This implies that

$$(d_{(v,\varphi(v))} F) \tau_{(v,\varphi(v))} V^s = \tau_{F(v,\varphi(v))} F(V^s). \tag{6.45}$$

Step 2. Invariant cones.

Given $\gamma \in (0, 1)$, consider the cones

$$C^s = \left\{ (v, w) \in E^s(x) \times E^u(x) : \|w\| < \gamma \|v\| \right\} \cup \{0\}$$

and

$$C^u = \left\{ (v, w) \in E^s(x) \times E^u(x) : \|v\| < \gamma \|w\| \right\} \cup \{0\}.$$

Lemma 6.2 *For any sufficiently small ε, given $(v, w) \in D_\varepsilon^s \times D_\varepsilon^u$, we have*

$$d_{(v,w)} F^{-1} C^s \subset C^s \quad and \quad d_{(v,w)} F C^u \subset C^u. \tag{6.46}$$

Proof Given $(y, z) \in E^s(x) \times E^u(x)$, let

$$(\bar{y}, \bar{z}) = d_{(v,w)} F(y, z)$$
$$= \left(Ay + d_{(v,w)} g(y, z), Bz + d_{(v,w)} h(y, z) \right). \tag{6.47}$$

We have

$$\|\bar{y}\| \le \tau \|y\| + K \|(y, z)\| \tag{6.48}$$

and

$$\|\bar{z}\| \ge \tau^{-1} \|z\| - K \|(y, z)\|. \tag{6.49}$$

For $(y, z) \in C^u \setminus \{0\}$, we also have $\|y\| < \gamma \|z\|$ and thus,

$$\|\bar{y}\| \le \tau \gamma \|z\| + K(1 + \gamma) \|z\|$$

and

$$\gamma \|\bar{z}\| \ge \tau^{-1} \gamma \|z\| - K \gamma (1 + \gamma) \|z\|.$$

Hence, taking ε sufficiently small so that

$$\tau\gamma + K(1+\gamma)^2 \leq \tau^{-1}\gamma,$$

we obtain $\|\bar{y}\| < \gamma\|\bar{z}\|$ and $(\bar{y}, \bar{z}) \in C^u$. This establishes the second inclusion in (6.46). The first inclusion can be obtained in a similar manner.   □

Step 3. Stable and unstable spaces.

Given $(v, w) \in D_\varepsilon^s \times D_\varepsilon^u$, consider the intersections

$$E^s(v, w) = \bigcap_{j=0}^{\infty} d_{(v,w)} F^{-j} \overline{C^s} \tag{6.50}$$

and

$$E^u(v, w) = \bigcap_{j=0}^{\infty} d_{(v,w)} F^j \overline{C^u}. \tag{6.51}$$

It follows from Lemma 6.2 that

$$E^s(v, w) \subset \overline{C^s} \quad \text{and} \quad E^u(v, w) \subset \overline{C^u}.$$

Lemma 6.3 *For any sufficiently small ε and γ, the sets $E^s(v, w)$ and $E^u(v, w)$ are subspaces of $T_x M$, respectively, of dimensions $\dim E^s(x)$ and $\dim E^u(x)$. Moreover, they vary continuously with (v, w) and*

$$E^s(v, w) \oplus E^u(v, w) = T_x M.$$

Proof We note that each set

$$H_j = d_{(v,w)} F^j \overline{C^u}$$

contains a subspace F_j^u of dimension $k = \dim E^u(x)$. Now let $v_{1j}, \ldots, v_{kj}$ be an orthonormal basis of F_j^u. Since H_j is nonincreasing in j (by Lemma 6.2) and the closed unit sphere of $E^u(x)$ is compact, there exists a sequence k_j such that

$$v_{ik_j} \to v_i \quad \text{when } j \to \infty,$$

for $i = 1, \ldots, k$, where $v_1, \ldots, v_k$ is some orthonormal set in $E^u(v, w)$. This shows that $E^u(v, w)$ contains a subspace G^u of dimension k. An analogous argument shows that $E^s(v, w)$ contains a subspace G^s of dimension $\dim M - k$. By (6.50) and (6.51), we have

$$G^s \subset E^s(v, w) \subset \overline{C^s} \quad \text{and} \quad G^u \subset E^u(v, w) \subset \overline{C^u}.$$

Since $\overline{C^s} \cap \overline{C^u} = \{0\}$, we obtain

$$G^s \oplus G^u = T_x M.$$

Now we show that the inclusions

$$G^s \subset E^s(v, w) \quad \text{and} \quad G^u \subset E^u(v, w)$$

are in fact equalities. Given

$$(y, z) \in E^s(x) \times E^u(x),$$

let $(\bar{y}, \bar{z})$ be as in (6.47). If $(y, z) \in \overline{C^u}$, then it follows from (6.48) and (6.49) that

$$\|(\bar{y}, \bar{z})\| \geq \|\bar{z}\| - \|\bar{y}\|$$
$$\geq \tau^{-1}\|z\| - K\|(y, z)\| - \tau\|y\| - K\|(y, z)\|.$$

Since $\|y\| \leq \gamma\|z\|$, we have

$$\|(y, z)\| \leq \|y\| + \|z\| \leq (1 + \gamma)\|z\|$$

and thus,

$$\|(\bar{y}, \bar{z})\| \geq (\tau^{-1} - \gamma\tau)\|z\| - 2K\|(y, z)\|$$
$$\geq \left(\frac{\tau^{-1} - \gamma\tau}{1 + \gamma} - 2K\right)\|(y, z)\|. \tag{6.52}$$

On the other hand, if $(y, z) \in \overline{C^s}$, then $\|z\| \leq \gamma\|y\|$ and

$$\|(\bar{y}, \bar{z})\| \leq \|\bar{y}\| + \|\bar{z}\|$$
$$\leq \tau\|y\| + K\|(y, z)\| + \tau^{-1}\|z\| + K\|(y, z)\|$$
$$\leq (\tau + \gamma\tau^{-1})\|y\| + 2K\|(y, z)\|.$$

Since

$$\|(y, z)\| \geq \|y\| - \|z\| \geq (1 - \gamma)\|y\|,$$

we obtain

$$\|(\bar{y}, \bar{z})\| \leq \left(\frac{\tau + \gamma\tau^{-1}}{1 - \gamma} + 2K\right)\|(y, z)\|. \tag{6.53}$$

Now we assume that $E^u(v, w) \setminus G^u \neq \emptyset$. Each vector $q \in E^u(v, w) \setminus G^u$ can be written in the form $q = q_s + q_u$, with $q_s \in G^s \setminus \{0\}$ and $q_u \in G^u$. It follows from (6.52) and (6.53) that

$$\|q_s\| \leq \left(\frac{\tau + \gamma\tau^{-1}}{1-\gamma} + 2K\right)^m \|d_{(v,w)}F^{-m}q_s\|$$

$$= \left(\frac{\tau + \gamma\tau^{-1}}{1-\gamma} + 2K\right)^m \|d_{(v,w)}F^{-m}(q - q_u)\|$$

$$\leq \alpha^m \|q - q_u\|, \tag{6.54}$$

where

$$\alpha := \frac{(\tau + \gamma\tau^{-1})/(1-\gamma) + 2K}{(\tau^{-1} - \gamma\tau)/(1+\gamma) - 2K}.$$

Taking ε and γ sufficiently small so that $\alpha < 1$ and letting $m \to \infty$ in (6.54), we obtain $q_s = 0$. This contradiction shows that $G^u = E^u(v, w)$. One can show in an analogous manner that $G^s = E^s(v, w)$.

Finally, we establish the continuous dependence of the spaces $E^s(v, w)$ and $E^u(v, w)$ on the point (v, w). By (6.52) and (6.53), we have

$$\|d_{(v,w)}F^m(y, z)\| \leq \left(\frac{\tau + \gamma\tau^{-1}}{1-\gamma} + 2K\right)^m \|(y, z)\|$$

for $(y, z) \in E^s(v, w)$ and $m > 0$, and

$$\|d_{(v,w)}F^{-m}(y, z)\| \leq \left(\frac{\tau^{-1} - \gamma\tau}{1+\gamma} - 2K\right)^m \|(y, z)\|$$

for $(y, z) \in E^u(v, w)$ and $m < 0$. One can now repeat the arguments in the proof of Theorem 5.1 to establish the continuity of the spaces $E^s(v, w)$ and $E^u(v, w)$ on (v, w). □

Step 4. C^1 regularity.

It remains to show that the function φ is of class C^1. Taking $\sigma < \gamma$, we have

$$\tau_{(v,\varphi(v))}V^s \subset \overline{C^s}. \tag{6.55}$$

On the other hand, it follows from (6.50) and (6.51) that

$$d_{(v,w)}FE^s(v, w) = E^s(F(v, w)) \tag{6.56}$$

and

$$d_{(v,w)}FE^u(v, w) = E^u(F(v, w)) \tag{6.57}$$

(compare with (5.30) and (5.31)). In particular, $E^s(v, w)$ and $E^u(v, w)$ are the largest sets contained, respectively, in $\overline{C^s}$ and $\overline{C^u}$ and satisfying (6.56) and (6.57). It follows from (6.45) and (6.55) that

$$\tau_{(v,\varphi(v))}V^s \subset E^s(v, \varphi(v))$$

for $v \in D_\varepsilon^s$. But since $\tau_{(v,\varphi(v))} V^s$ and $E^s(v, \varphi(v))$ project both onto D_ε^s (we recall that V^s is a graph over D_ε^s), we obtain

$$\tau_{(v,\varphi(v))} V^s = E^s\big(v, \varphi(v)\big) \tag{6.58}$$

and thus, $\tau_{(v,\varphi(v))} V^s$ is a linear space of dimension

$$\dim E^s\big(v, \varphi(v)\big) = \dim E^s(x).$$

It follows from Lemma 6.1 that the function φ is differentiable. Moreover, by Lemma 6.3, the function $v \mapsto E^s(v, \varphi(v))$ is continuous since it is a composition of continuous functions and thus, V^s is a manifold of class C^1. One can show in an analogous manner that V^u is a manifold of class C^1. Taking $v = 0$, it follows from (6.58) and the corresponding identity for V^u that

$$T_0 V^s = E^s \quad \text{and} \quad T_0 V^u = E^u.$$

This establishes the identities in (6.37).   □

Definition 6.2 The manifolds $V^s(x)$ and $V^u(x)$ or, more precisely, the manifolds $V^s(x) \cap B$ and $V^u(x) \cap B$ are called, respectively, *stable* and *unstable manifolds* at the point x.

By (6.36), $V^s(x)$ is forward f-invariant and $V^u(x)$ is backward f-invariant.

Theorem 6.2 is a particular case of a more general result (Theorem 6.3) for arbitrary hyperbolic sets.

6.2 Stable and Unstable Invariant Manifolds

In this section we establish a basic but also fundamental result on the behavior of the orbits of a diffeomorphism with a hyperbolic set. It is in fact a substantial generalization of the Hadamard–Perron theorem (Theorem 6.2) on the existence of stable and unstable manifolds for a hyperbolic fixed point. More precisely, it establishes the existence of stable and unstable manifolds for all points of a hyperbolic set.

6.2.1 Existence of Invariant Manifolds

Let Λ be a hyperbolic set for a C^1 diffeomorphism $f : \mathbb{R}^p \to \mathbb{R}^p$. Given $\varepsilon > 0$, for each $x \in \Lambda$, we consider the sets

$$V^s(x) = \big\{ y \in B(x, \varepsilon) : \big\| f^n(y) - f^n(x) \big\| < \varepsilon \text{ for } n > 0 \big\} \tag{6.59}$$

and

$$V^u(x) = \{y \in B(x, \varepsilon) : \|f^n(y) - f^n(x)\| < \varepsilon \text{ for } n < 0\}, \qquad (6.60)$$

where $B(x, \varepsilon) \subset \mathbb{R}^p$ is the ball of radius ε centered at x. Clearly, $x \in V^s(x) \cap V^u(x)$. We also have

$$f(V^s(x)) \subset V^s(f(x)) \quad \text{and} \quad f^{-1}(V^u(x)) \subset V^u(f^{-1}(x)). \qquad (6.61)$$

Example 6.1 Let Λ be the Smale horseshoe constructed in Sect. 5.2.2. For any point $x \in \Lambda$ outside the boundary of the square $[0, 1]^2$, given $\varepsilon > 0$ sufficiently small, the set $V^s(x)$ is a horizontal line segment and the set $V^u(x)$ is a vertical line segment. In order to determine $V^s(x)$ and $V^u(x)$ at the remaining points, we would need to know explicitly the diffeomorphism f outside the square.

Now we consider arbitrary hyperbolic sets.

Theorem 6.3 (Stable and Unstable Manifolds) *Let Λ be a hyperbolic set for a C^1 diffeomorphism $f : \mathbb{R}^p \to \mathbb{R}^p$. For any sufficiently small $\varepsilon > 0$, the following properties hold:*

1. *for each $x \in \Lambda$, the sets $V^s(x)$ and $V^u(x)$ are manifolds of class C^1 satisfying*

$$T_x V^s(x) = E^s(x) \quad \text{and} \quad T_x V^u(x) = E^u(x);$$

2. *there exist $\rho \in (0, 1)$ and $C > 0$ such that*

$$\|f^n(y) - f^n(x)\| \le C\rho^n \|y - x\| \quad \text{for } y \in V^s(x)$$

and

$$\|f^{-n}(y) - f^{-n}(x)\| \le C\rho^n \|y - x\| \quad \text{for } y \in V^u(x),$$

for any $x \in \Lambda$ and $n \in \mathbb{N}$.

Proof The proof is an elaboration of the proof of Theorem 6.2 and so we only describe the changes that are necessary.

For each $x \in \Lambda$, consider the map $f_x : \mathbb{R}^p \to \mathbb{R}^p$ defined by

$$f_x(y) = f(y + x) - f(x).$$

One can write f_x in the form

$$f_x(v, w) = \left(A_x^s v + g_x^s(v, w), A_x^u w + g_x^u(v, w)\right) \in E^s(x) \oplus E^u(x)$$

for each $(v, w) \in E^s(x) \oplus E^u(x)$, where

$$A_x^s = d_x f | E^s(x) : E^s(x) \to E^s(f(x))$$

and

$$A_x^u = d_x f | E^u(x) \colon E^u(x) \to E^u \big(f(x) \big).$$

We note that A_x^s and A_x^u are invertible linear transformations and that g_x^s and g_x^u are C^1 functions with

$$g_x^s(0) = 0, \qquad g_x^u(0) = 0, \qquad d_0 g_x^s = 0, \qquad d_0 g_x^u = 0.$$

Moreover, one can always assume that the inner products $\langle \cdot, \cdot \rangle_x$ in the tangent spaces $T_x \mathbb{R}^p$ are such that

$$\big\| A_x^s \big\| < \tau \quad \text{and} \quad \big\| (A_x^u)^{-1} \big\| < \tau$$

for any $x \in \Lambda$ and some constant $\tau \in (0, 1)$.

Now we consider separately each orbit of f. Given $x \in \Lambda$ and $n \in \mathbb{Z}$, write the map $F_n = f_{f^n(x)}$ in the form

$$F_n(v, w) = \big(A_n v + g_n(v, w), B_n w + h_n(v, w) \big), \tag{6.62}$$

where

$$A_n = A_{f^n(x)}^s, \qquad B_n = A_{f^n(x)}^u, \qquad g_n = g_{f^n(x)}^s, \qquad h_n = g_{f^n(x)}^u.$$

We also consider the spaces

$$E_n^s = E^s \big(f^n(x) \big) \quad \text{and} \quad E_n^u = E^u \big(f^n(x) \big).$$

The linear transformations

$$A_n \colon E_n^s \to E_{n+1}^s \quad \text{and} \quad B_n \colon E_n^u \to E_{n+1}^u$$

satisfy

$$\| A_n \| < \tau \quad \text{and} \quad \big\| B_n^{-1} \big\| < \tau$$

for each $n \in \mathbb{Z}$. Moreover,

$$\big(g_x^s, g_x^u \big)(y) = f(y + x) - f(x) - d_x f y$$

and the function

$$G \colon (x, y) \mapsto d_y \big(g_x^s, g_x^u \big) = d_{y+x} f - d_x f$$

is continuous since f is of class C^1. Since Λ is compact, given $\varepsilon > 0$ and open balls

$$B_{2\varepsilon}^s(x) \subset E^s(x) \quad \text{and} \quad B_{2\varepsilon}^u(x) \subset E^u(x)$$

of radius 2ε centered at the origin, the function G is uniformly continuous on the set

$$\big\{ (x, y) : x \in \Lambda, \, y \in D_\varepsilon(x) \big\}, \quad \text{where } D_\varepsilon(x) = \overline{B_{2\varepsilon}^s(x)} \oplus \overline{B_{2\varepsilon}^u(x)}.$$

Since $G(x, 0) = 0$, we have

$$K := \sup\{\|G(x, y)\| : (x, y) \in \Lambda \times D_\varepsilon(x)\} \to 0$$

when $\varepsilon \to 0$. Moreover,

$$\|d_{(v,w)}g_n\| \leq K \quad \text{and} \quad \|d_{(v,w)}h_n\| \leq K$$

for $n \in \mathbb{Z}$ and $(v, w) \in \overline{D_\varepsilon^s} \oplus \overline{D_\varepsilon^u}$, where

$$D_\varepsilon^s = B_\varepsilon^s\big(f^n(x)\big) \quad \text{and} \quad D_\varepsilon^u = B_\varepsilon^u\big(f^n(x)\big).$$

Given $\sigma \in (0, 1]$, let X be the space of sequences $\varphi = (\varphi_n)_{n \in \mathbb{Z}}$ of functions $\varphi_n \colon D_\varepsilon^s \to E_n^u$ such that $\varphi_n(0) = 0$ and

$$\|\varphi_n(v) - \varphi_n(w)\| \leq \sigma \|v - w\|$$

for $n \in \mathbb{Z}$ and $v, w \in D_\varepsilon^s$. One can easily verify that X is a complete metric space with the distance

$$d(\varphi, \psi) = \sup\{\|\varphi_n(v) - \psi_n(v)\| : n \in \mathbb{Z}; \ v \in D_\varepsilon^s\}.$$

We want to show that there exist functions φ_n such that

$$V^s\big(f^n(x)\big) \supset x + \{(v, \varphi_n(v)) : v \in D_\varepsilon^s\} \qquad (6.63)$$

for $n \in \mathbb{Z}$. We first proceed formally, assuming that (6.63) holds. It follows from (6.62) and the invariance in (6.61) that

$$F_n\big(v, \varphi_n(v)\big) = \big(A_n v + g_n\big(v, \varphi_n(v)\big), B_n \varphi_n(v) + h_n\big(v, \varphi_n(v)\big)\big)$$

and

$$B_n \varphi_n(v) + h_n\big(v, \varphi_n(v)\big) = \varphi_{n+1}\big(A_n v + g_n\big(v, \varphi_n(v)\big)\big).$$

Thus,

$$\varphi_n(v) = B_n^{-1} \varphi_{n+1}\big(A_n v + g_n\big(v, \varphi_n(v)\big)\big) - B_n^{-1} h_n\big(v, \varphi_n(v)\big)$$

for each $n \in \mathbb{Z}$. This leads us to introduce a map T on X by

$$T(\varphi)_n(v) = B_n^{-1} \varphi_{n+1}\big(A_n v + g_n\big(v, \varphi_n(v)\big)\big) - B_n^{-1} h_n\big(v, \varphi_n(v)\big).$$

We note that $\varphi = (\varphi_n)_{n \in \mathbb{Z}}$ is a fixed point of T if and only if the graphs of the functions φ_n in (6.63) satisfy the invariance property in (6.61). Hence, if we prove that T has a unique fixed point $\varphi \in X$, then the sets on the right-hand side of (6.63) coincide with $V^s(f^n(x))$ in some open neighborhood of $f^n(x)$.

The remaining steps are analogous to those in the proof of Theorem 6.2 and thus are omitted. $\qquad\qquad\qquad\qquad\qquad\qquad\qquad\qquad\qquad\qquad\qquad\qquad\qquad\quad$ $\square$

6.2.2 Local Product Structure

In this section we show that some hyperbolic sets have what is called a local product structure. Let Λ be a hyperbolic set for a C^1 diffeomorphism $f \colon \mathbb{R}^p \to \mathbb{R}^p$. Given $x, y \in \Lambda$, we write

$$[x, y] = V^s(x) \cap V^u(y).$$

Definition 6.3 Λ is said to have a *local product structure* if there exist $\varepsilon > 0$ and $\delta > 0$ such that

$$\mathrm{card}[x, y] = 1 \quad \text{and} \quad [x, y] \in \Lambda$$

for any $x, y \in \Lambda$ with $\|x - y\| < \delta$ (with the constant ε as in Theorem 6.3).

When Λ has a local product structure, we obtain a function

$$[\cdot, \cdot] \colon \big\{(x, y) \in \Lambda \times \Lambda : \|x - y\| < \delta\big\} \to \Lambda.$$

Now we consider a class of hyperbolic sets for which there exists a local product structure.

Definition 6.4 A hyperbolic set Λ for a diffeomorphism $f \colon \mathbb{R}^p \to \mathbb{R}^p$ is said to be *locally maximal* if there exists an open neighborhood $U \supset \Lambda$ such that

$$\Lambda = \bigcap_{n \in \mathbb{Z}} f^n(U).$$

In other words, a hyperbolic set is locally maximal if all orbits remaining in some open neighborhood of Λ are in fact in Λ.

Theorem 6.4 *Any locally maximal hyperbolic set Λ for a C^1 diffeomorphism f has a local product structure.*

Proof By Theorem 5.1, the spaces $E^s(x)$ and $E^u(x)$ vary continuously with $x \in \Lambda$. This implies that the map

$$\Lambda \ni x \mapsto \angle\big(E^s(x), E^u(x)\big) \in (0, \pi/2]$$

is continuous since it is a composition of continuous functions. On the other hand, since Λ is compact, there exists an $\alpha > 0$ such that

$$\angle\big(E^s(x), E^u(x)\big) > \alpha \quad \text{for } x \in \Lambda.$$

Now we show that there exists a $\delta > 0$ such that

$$\angle\big(E^s(x), E^u(y)\big) > \frac{\alpha}{2} \tag{6.64}$$

for any $x, y \in \Lambda$ with $\|x - y\| \leq \delta$. Given $x \in \Lambda$, let $U_x \subset \Lambda \times \Lambda$ be an open ball centered at the pair (x, x) such that

$$\angle\left(E^s(y), E^u(z)\right) > \frac{\alpha}{2} \quad \text{for } (y, z) \in U_x. \tag{6.65}$$

This is always possible since the function $(y, z) \mapsto (E^s(y), E^u(z))$ is continuous. Moreover, since Λ is compact, the diagonal

$$D = \{(x, x) : x \in \Lambda\} \subset \Lambda \times \Lambda$$

is also compact and there exists a finite subcover $U_{x_1}, \ldots, U_{x_n}$ of D. Now take $\delta > 0$ such that any open ball of radius 2δ centered at a point $(x, x) \in \Lambda \times \Lambda$ is contained in some open set U_{x_i} (it is sufficient to assume that 2δ is a Lebesgue number[2] of the open cover of D formed by the sets $U_{x_i} \cap D$, for $i = 1, \ldots, n$). Given $x, y \in \Lambda$ with $\|x - y\| \leq \delta$, we have

$$\|(x, y) - (x, x)\| = \|(0, y - x)\| \leq \|y - x\| \leq \delta.$$

Hence, the point (x, y) is in some open set U_{x_i} and it follows from (6.65) that inequality (6.64) holds.

Now take $\varepsilon > 0$ in (6.59) and (6.60) such that

$$\bigcup_{x \in \Lambda} \left(V^s(x) \cup V^u(x)\right) \subset U, \tag{6.66}$$

with the open set U as in Definition 6.4. By Theorem 6.3, the manifolds $V^s(x)$ and $V^u(x)$ are graphs of C^1 functions. Moreover, they are tangent, respectively, to the spaces $E^s(x)$ and $E^u(x)$. This implies that taking ε sufficiently small, one can make the constant σ in (6.41) as small as desired. Since the angle between $E^s(x)$ and each tangent space to $V^s(x)$ is at most $\tan^{-1}\sigma$, the angle between $E^u(y)$ and each tangent space to $V^s(x)$ is contained in the interval

$$\left(\frac{\alpha}{2} - \tan^{-1}\sigma, \frac{\alpha}{2} + \tan^{-1}\sigma\right)$$

(see Fig. 6.2). Thus, the angle between each tangent space to $V^s(x)$ and each tangent space to $V^u(y)$ is contained in the interval

$$\left(\frac{\alpha}{2} - 2\tan^{-1}\sigma, \frac{\alpha}{2} + 2\tan^{-1}\sigma\right)$$

(see Fig. 6.2). Taking σ sufficiently small so that

$$\frac{\alpha}{2} - 2\tan^{-1}\sigma > 0 \quad \text{and} \quad \frac{\alpha}{2} + 2\tan^{-1}\sigma < \frac{\pi}{2},$$

---

[2]**Theorem** (See for example [43]) Given an open cover of a compact metric space X, there exists a positive number δ (called a Lebesgue number of the cover) such that any subset of X of diameter less than δ is contained in some element of the cover.

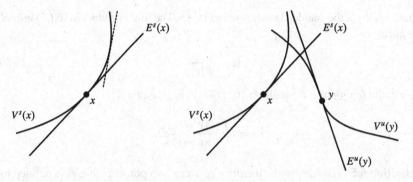

Fig. 6.2 Tangent spaces to the manifolds $V^s(x)$ and $V^u(y)$

we obtain

$$\text{card}[x, y] = \text{card}\big(V^s(x) \cap V^u(y)\big) \leq 1$$

for any $x, y \in \Lambda$ with $\|x - y\| \leq \delta$. Since the sizes of the open neighborhoods D_ε^s and D_ε^u (see (6.40) and (6.42)) only depend on ε, taking δ sufficiently small yields that $\text{card}[x, y] = 1$ for any $x, y \in \Lambda$ with $\|x - y\| < \delta$.

Finally, it follows from (6.61) and (6.66) that $f^n([x, y]) \in U$ for $n \in \mathbb{Z}$. Since the set Λ is locally maximal, we conclude that $[x, y] \in \Lambda$. $\square$

6.3 Geodesic Flows

In this section we consider the geodesic flow in hyperbolic geometry. In particular, we show how it gives rise to examples of hyperbolic flows. More precisely, after a brief introduction to hyperbolic geometry and its geodesic flow, we introduce the notion of a hyperbolic set for a flow and we show how to obtain examples of hyperbolic sets from the geodesic flow. We note that some basic results of hyperbolic geometry are formulated without proof.

6.3.1 Hyperbolic Geometry

Consider the upper half-plane

$$\mathbb{H} = \big\{z \in \mathbb{C} : \text{Im}\, z > 0\big\},$$

with the inner product in each tangent space $T_z\mathbb{H} = \mathbb{C} = \mathbb{R}^2$ given by

$$\langle v, w \rangle_z = \frac{\langle v, w \rangle}{(\text{Im}\, z)^2}, \tag{6.67}$$

where $\langle v, w \rangle$ is the standard inner product in $\mathbb{R}^2$. The inner product in (6.67) induces the norm

$$|v|_z = \frac{|v|}{\operatorname{Im} z}. \tag{6.68}$$

Hence, the *length* of a C^1 path $\gamma : [0, \tau] \to \mathbb{H}$ is given by

$$L_\gamma = \int_0^\tau \frac{|\gamma'(t)|}{\operatorname{Im} \gamma(t)} \, dt.$$

Definition 6.5 The *hyperbolic distance* between two points $z, w \in \mathbb{H}$ is defined by

$$d(z, w) = \inf L_\gamma,$$

where the infimum is taken over all C^1 paths joining z to w, that is, all C^1 paths $\gamma : [0, \tau] \to \mathbb{H}$ with $\gamma(0) = z$ and $\gamma(\tau) = w$.

Now let $S^*L(2, \mathbb{R})$ be the group of matrices

$$A = \begin{pmatrix} a & b \\ c & d \end{pmatrix}$$

with real entries and determinant 1 or -1, and define maps T_A on $\mathbb{H}$ by

$$T_A(z) = \frac{az + b}{cz + d} \quad \text{or} \quad T_A(z) = \frac{a\bar{z} + b}{c\bar{z} + d}, \tag{6.69}$$

respectively, when $ad - bc = 1$ or $ad - bc = -1$. Clearly, the matrices A and $-A$ represent the same map, that is, $T_{-A} = T_A$.

We also recall the notion of an isometry.

Definition 6.6 A map $T : \mathbb{H} \to \mathbb{H}$ is said to be an *isometry* if

$$d\big(T(z), T(w)\big) = d(z, w)$$

for any $z, w \in \mathbb{H}$.

Now we show that the elements of the group $S^*L(2, \mathbb{R})$ or, more precisely, the maps in (6.69), are isometries.

Proposition 6.3 *The maps T_A in (6.69) take $\mathbb{H}$ to itself and are isometries.*

Proof We first show that $S^*L(2, \mathbb{R})$ is generated by the group $SL(2, \mathbb{R})$ of matrices

$$A = \begin{pmatrix} a & b \\ c & d \end{pmatrix}$$

with real entries and determinant 1 and by the matrix

$$B = \begin{pmatrix} -1 & 0 \\ 0 & 1 \end{pmatrix}.$$

Indeed, if A has determinant 1, then the matrix

$$AB = \begin{pmatrix} a & b \\ c & d \end{pmatrix} \begin{pmatrix} -1 & 0 \\ 0 & 1 \end{pmatrix} = \begin{pmatrix} -a & b \\ -c & d \end{pmatrix}$$

has determinant -1. This yields the desired result since all matrices with determinant -1 can be written in this form.

Now we show that the maps T_A defined by matrices A with determinant 1 (called Möbius transformations) and the map $z \mapsto -\overline{z} = T_B(z)$ take $\mathbb{H}$ to itself. If

$$w = T_A(z) = \frac{az+b}{cz+d},$$

then

$$w = \frac{(az+b)(c\overline{z}+d)}{|cz+d|^2}$$

$$= \frac{ac|z|^2 + adz + bc\overline{z} + bd}{|cz+d|^2}$$

and thus,

$$\operatorname{Im} w = \frac{w - \overline{w}}{2i}$$

$$= \frac{(ad-bc)z - (ad-bc)\overline{z}}{2i|cz+d|^2}$$

$$= \frac{z - \overline{z}}{2i|cz+d|^2}$$

$$= \frac{\operatorname{Im} z}{|cz+d|^2} > 0. \tag{6.70}$$

This shows that $w \in \mathbb{H}$. We also have

$$\operatorname{Im}(-\overline{z}) = \frac{-\overline{z} - \overline{-\overline{z}}}{2i}$$

$$= \frac{z - \overline{z}}{2i} = \operatorname{Im} z > 0.$$

It remains to show that these maps preserve the hyperbolic distance. For this it is sufficient to show that the length of any C^1 path is preserved under their action.

Indeed, the image of a C^1 path under a map T_A is still a C^1 path and all C^1 paths are of this form. Hence,

$$d\big(T_A(z), T_A(w)\big) = \inf L_\alpha,$$

where the infimum is taken over all paths $\alpha = T_A \circ \gamma$ obtained from a C^1 path γ joining z to w. Now let $\gamma : [0, \tau] \to \mathbb{H}$ be a C^1 path and let

$$\alpha(t) = T_A(\gamma(t)) = \frac{a\gamma(t)+b}{c\gamma(t)+d},$$

with $A \in SL(2, \mathbb{R})$. We have

$$\begin{aligned}
\alpha'(t) &= \frac{a\gamma'(t)(c\gamma(t)+d) - c\gamma'(t)(a\gamma(t)+b)}{(c\gamma(t)+d)^2} \\
&= \frac{ad-bc}{(c\gamma(t)+d)^2}\gamma'(t) \\
&= \frac{1}{(c\gamma(t)+d)^2}\gamma'(t)
\end{aligned} \tag{6.71}$$

and, as in (6.70),

$$\operatorname{Im}\alpha(t) = \frac{\operatorname{Im}\gamma(t)}{|c\gamma(t)+d|^2}.$$

Hence,

$$\begin{aligned}
L_\alpha &= \int_0^\tau \frac{|\alpha'(t)|}{\operatorname{Im}\alpha(t)}\,dt \\
&= \int_0^\tau \frac{|\gamma'(t)|}{|c\gamma(t)+d|^2} \cdot \frac{|c\gamma(t)+d|^2}{\operatorname{Im}\gamma(t)}\,dt \\
&= \int_0^\tau \frac{|\gamma'(t)|}{\operatorname{Im}\gamma(t)}\,dt = L_\gamma.
\end{aligned}$$

Finally, if $\alpha(t) = -\overline{\gamma(t)}$, then

$$\alpha'(t) = -\overline{\gamma'(t)} \quad \text{and} \quad \operatorname{Im}\alpha(t) = \operatorname{Im}\gamma(t).$$

Hence,

$$L_\alpha = \int_0^\tau \frac{|\alpha'(t)|}{\operatorname{Im}\alpha(t)}\,dt = \int_0^\tau \frac{|\gamma'(t)|}{\operatorname{Im}\gamma(t)}\,dt = L_\gamma.$$

This completes the proof of the proposition.   □

Since the matrices

$$\begin{pmatrix} a & b \\ c & d \end{pmatrix} \quad \text{and} \quad -\begin{pmatrix} a & b \\ c & d \end{pmatrix}$$

Fig. 6.3 Geodesics of $\mathbb{H}$

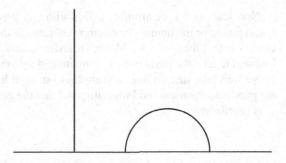

represent the same map, we also consider the groups

$$PSL(2, \mathbb{R}) = SL(2, \mathbb{R})/\{\mathrm{Id}, -\mathrm{Id}\}$$

and

$$PS^*L(2, \mathbb{R}) = S^*L(2, \mathbb{R})/\{\mathrm{Id}, -\mathrm{Id}\},$$

where two matrices are identified if one is the negative of the other. Proposition 6.3 says that $PS^*L(2, \mathbb{R})$ and its subgroup $PSL(2, \mathbb{R})$ are formed by isometries.

Now we consider the geodesics, that is, the shortest paths between two points.

Proposition 6.4 *The geodesics between two points of $\mathbb{H}$ are the vertical half-lines and the semicircles centered at points of the real axis (see Fig. 6.3). More precisely, given $z \in \mathbb{H}$ and $v \in \mathbb{C} \setminus \{0\}$:*

1. *if v is parallel to the imaginary axis, then the geodesic passing through z with direction v at this point is the half-line $\{z + vt : t \in \mathbb{R}\} \cap \mathbb{H}$;*
2. *if v is not parallel to the imaginary axis, then the geodesic passing through z with direction v at this point is the semicircle centered at the real axis that is tangent to v at the point z.*

Proof Take $z = ic$ and $w = id$, with $d > c$. Let $\gamma : [0, \tau] \to \mathbb{H}$ be a C^1 path with $\gamma(0) = z$ and $\gamma(\tau) = w$. Writing $\gamma(t) = x(t) + iy(t)$, we obtain

$$L_\gamma = \int_0^\tau \frac{|\gamma'(t)|}{y(t)}\, dt \geq \int_0^\tau \frac{|y'(t)|}{y(t)}\, dt$$

$$\geq \int_0^\tau \frac{y'(t)}{y(t)}\, dt = \log y(t)|_{t=0}^{t=\tau} = \log \frac{d}{c}. \tag{6.72}$$

On the other hand, for the C^1 path $\alpha : [c, d] \to \mathbb{H}$ defined by $\alpha(t) = it$, we have

$$L_\alpha = \int_c^d \frac{|\alpha'(t)|}{\operatorname{Im}\alpha(t)}\, dt = \int_c^d \frac{1}{t}\, dt = \log \frac{d}{c}.$$

Comparing with (6.72), we conclude that the geodesic joining ic to id is the vertical line segment between these two points.

Now let $z, w \in \mathbb{H}$ be arbitrary points with $z \neq w$ and let R be the unique vertical line segment or the unique arc of circle centered at the real axis joining z to w. One can show that there exists a Möbius transformation T such that $T(R)$ is a vertical line segment on the positive part of the imaginary axis (we recall that Möbius transformations take straight lines and circles to straight lines or circles). It follows from the previous argument and Proposition 6.3 that the geodesic joining the points z and w is precisely R. $\square$

6.3.2 Quotients by Isometries

In this section we consider the quotient of $\mathbb{H}$ by subgroups of isometries. The procedure is analogous to an alternative construction of the torus $\mathbb{T}^n$. Namely, let $T_i \colon \mathbb{R}^n \to \mathbb{R}^n$, for $i = 1, \ldots, n$, be the translations

$$T_i(x) = x + e_i,$$

where e_i is the ith vector of the standard basis of $\mathbb{R}^n$. Then the torus $\mathbb{T}^n$ is obtained identifying the points of $\mathbb{R}^n$ that can be obtained from each other applying successively any of the maps T_i and their inverses T_i^{-1}. Given $m = (m_1, \ldots, m_n) \in \mathbb{Z}$, we have

$$\left(T_1^{m_1} \circ T_2^{m_2} \circ \cdots \circ T_n^{m_n}\right)(x) = x + m.$$

Since the maps $T_1, \ldots, T_n$ commute, the points $x + m$ are precisely those that can be obtained from x successively applying the maps T_i and their inverses. In other words, two points $x, y \in \mathbb{R}^n$ are identified if and only if $x - y \in \mathbb{Z}^n$ and thus, this procedure yields the torus $\mathbb{T}^n$.

The *genus* of a connected orientable surface M is the largest number g of closed simple curves $\gamma_1, \ldots, \gamma_g \subset M$ with

$$\gamma_j \cap \gamma_j = \varnothing \quad \text{for } i \neq j$$

such that $M \setminus \bigcup_{i=1}^g \gamma_i$ is connected. It can be described as the number of handles on the surface. For example, the sphere has genus 0 and the torus $\mathbb{T}^2$ has genus 1 (see Fig. 6.4). See Fig. 6.5 for a compact surface of genus 2.

Now we consider the group of isometries $PSL(2, \mathbb{R})$. We show that each compact surface of genus at least 2 can be obtained as the quotient of $\mathbb{H}$ by a subgroup of $PSL(2, \mathbb{R})$.

Proposition 6.5 *Given $g \geq 2$, there exists a subgroup G of $PSL(2, \mathbb{R})$ such that the quotient $\mathbb{H}/G$ is a compact surface of genus g.*

Proof We first consider the map

$$T(z) = \frac{z - i}{z + i}.$$

Fig. 6.4 The torus $\mathbb{T}^2$

Fig. 6.5 A compact surface
of genus 2

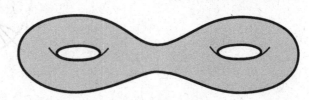

One can easily verify that T maps $\mathbb{H}$ onto the unit disk

$$\mathbb{D} = \{z \in \mathbb{C} : |z| < 1\}.$$

Indeed, $T(i) = 0$ and each $x \in \mathbb{R}$ is mapped to a point with modulus

$$|T(x)| = \left|\frac{x-i}{x+i}\right| = \frac{|x-i|}{|x+i|} = 1.$$

Moreover, the map T takes geodesics to diameters of $\mathbb{D}$ or circular arcs that are orthogonal to the boundary of $\mathbb{D}$.

Given $r \in (0, 1)$, consider the points $z_k = re^{i\pi k/4}$ for $k = 0, \ldots, 7$. We also consider the circular arcs

$$R_1, \ R_2, \ R_3, \ R_4, \ R_1', \ R_2', \ R_3', \ R_4' \tag{6.73}$$

that are orthogonal to the boundary of $\mathbb{D}$, determined successively by the pairs of points

$$(z_0, z_1), \ (z_1, z_2), \ \ldots, \ (z_6, z_7), \ (z_7, z_0)$$

(see Fig. 6.6). Now take r such that the sum of the interior angles of the octagon in Fig. 6.6 is 2π. One can show that there exist unique Möbius transformations T_j, for $j = 1, 2, 3, 4$, such that

$$S_j(R_j) = R_j', \quad \text{where } S_j = T \circ T_j \circ T^{-1},$$

reversing the direction of the arcs, that is,

$$S_1(z_0) = z_5, \qquad S_2(z_1) = z_6, \qquad S_3(z_2) = z_7, \qquad S_4(z_3) = z_0.$$

Finally, let G be the group generated by the Möbius transformations T_1, T_2, T_3 and T_4. One can verify that the quotient $\mathbb{H}/G$ is a compact surface of genus 2.

Replacing the 8 arcs in (6.73) by $4g$ arcs such that the sum of the interior angles of the polygon that they determine is equal to 2π, we obtain in an analogous manner a quotient of $\mathbb{H}$ that is a compact surface of genus g. $\square$

Fig. 6.6 Arcs R_j and R'_j for $j = 1, 2, 3, 4$

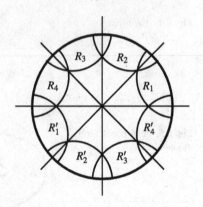

6.3.3 Geodesic Flow

In this section we describe the geodesic flow on $\mathbb{H}$ or, more precisely, on its unit tangent bundle

$$S\mathbb{H} = \big\{ (z, v) \in \mathbb{H} \times \mathbb{C} : |v|_z = 1 \big\},$$

where the norm $|v|_z$ is given by (6.68).

Example 6.2 The C^1 path $\gamma \colon \mathbb{R} \to \mathbb{H}$ defined by $\gamma(t) = ie^t$ travels along the geodesic $\{ z \in \mathbb{H} : \operatorname{Re} z = 0 \}$. Moreover,

$$
\begin{aligned}
|\gamma'(t)|_{\gamma(t)} &= \frac{\langle ie^t, ie^t \rangle^{1/2}}{\operatorname{Im}(ie^t)} \\
&= \frac{\langle (0, e^t), (0, e^t) \rangle^{1/2}}{e^t} = 1.
\end{aligned}
$$

Thus, we obtain a path

$$\mathbb{R} \ni t \mapsto \big(\gamma(t), \gamma'(t) \big) \in S\mathbb{H}$$

in the unit tangent bundle with $(\gamma(0), \gamma'(0)) = (i, i)$.

Now let us take $(z, v) \in S\mathbb{H}$. One can show that there exists a unique Möbius transformation T such that (see Fig. 6.7)

$$T(i) = z \quad \text{and} \quad T'(i)i = v,$$

which thus takes the geodesic ie^t traversing the positive part of the imaginary axis to the geodesic $\gamma(t)$ passing through z with direction v at this point. More precisely, let $x, y \in \mathbb{R} \cup \{\infty\}$ be, respectively, the limits $\gamma(-\infty)$ and $\gamma(+\infty)$. We consider four cases:

Fig. 6.7 The Möbius transformation T

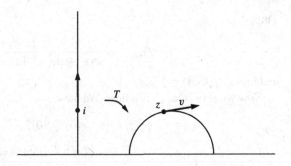

1. when $x, y \in \mathbb{R}$ and $x < y$, we have

$$T(w) = \frac{\alpha y w + x}{\alpha w + 1}, \quad \text{where } \alpha = \left| \frac{z - x}{z - y} \right|;$$

2. when $x, y \in \mathbb{R}$ and $x > y$, we have

$$T(w) = \frac{y w - \alpha x}{w - \alpha}, \quad \text{where } \alpha = \left| \frac{z - y}{z - x} \right|;$$

3. when $x \in \mathbb{R}$ and $y = \infty$, we have

$$T(w) = \alpha w + x, \quad \text{where } \alpha = \operatorname{Im} z;$$

4. when $x = \infty$ and $y \in \mathbb{R}$, we have

$$T(w) = -\alpha / w + y, \quad \text{where } \alpha = \operatorname{Im} z.$$

Now we use the map T (that depends on z and v) to introduce the geodesic flow.

Definition 6.7 The *geodesic flow* $\varphi_t \colon S\mathbb{H} \to S\mathbb{H}$ is defined by

$$\varphi_t(z, v) = \big(\gamma(t), \gamma'(t)\big),$$

where $\gamma(t) = T(ie^t)$.

We verify that φ_t is indeed a flow.

Proposition 6.6 *The family of maps φ_t is a flow of $S\mathbb{H}$.*

Proof Writing

$$T(z) = \frac{az + b}{cz + d},$$

it follows from (6.70) and (6.71) that

$$\operatorname{Im} \gamma(t) = \frac{\operatorname{Im}(ie^t)}{|cie^t + d|^2} \quad \text{and} \quad \gamma'(t) = \frac{ie^t}{(cie^t + d)^2}.$$

Thus,

$$|\gamma'(t)|_{\gamma(t)} = \frac{e^t}{|cie^t + d|^2} \cdot \frac{|cie^t + d|^2}{\operatorname{Im}(ie^t)} = 1$$

and hence, $\varphi_t(S\mathbb{H}) \subset S\mathbb{H}$.

Now we show that φ_t is a flow. We have

$$\varphi_0(z, v) = (\gamma(0), \gamma'(0))$$
$$= (T(i), T'(i)i) = (z, v).$$

Moreover,

$$(\varphi_t \circ \varphi_s)(z, v) = \varphi_t(\gamma(s), \gamma'(s))$$

with $\gamma(s) = S^{-1}(ie^s)$, where $S(w) = T(e^s w)$. Indeed, writing $R = S^{-1}$, we have

$$R(\gamma(s)) = e^{-s}T^{-1}(T(ie^s)) = i$$

and

$$R'(\gamma(s))\gamma'(s) = (R \circ \gamma)'(t)|_{t=s}$$
$$= (ie^{-s+t})'\big|_{t=s} = i.$$

Hence,

$$(\varphi_t \circ \varphi_s)(z, v) = (\alpha(t), \alpha'(t)),$$

where

$$\alpha(t) = S(ie^{it}) = T(ie^{t+s}) = \gamma(t+s),$$

which yields the identity

$$(\varphi_t \circ \varphi_s)(z, v) = (\gamma(t+s), \gamma'(t+s)) = \varphi_{t+s}(z, v).$$

This completes the proof of the proposition.   □

We also introduce a distance on $S\mathbb{H}$. Given $(z, v), (z', v') \in S\mathbb{H}$, let $\gamma \colon [0, 1] \to \mathbb{H}$ be the unique geodesic arc such that

$$\gamma(0) = z \quad \text{and} \quad \gamma(1) = z'.$$

Let also $F \colon [0, 1] \to \mathbb{H} \setminus \{0\}$ be a continuous vector field with $F(0) = v$ such that the angle between $F(t)$ and $\gamma'(t)$ is equal to the angle between v and $\gamma'(0)$ for every $t \in [0, 1]$. The distance between (z, v) and (z', v') is defined by

$$d((z, v), (z', v')) = \sqrt{d(z, z')^2 + \alpha^2}, \tag{6.74}$$

where α is the angle between $F(1)$ and v'. One can verify that there exist inner products $\langle \cdot, \cdot \rangle_{(z,v)}$ on the tangent spaces $T_{(z,v)}S\mathbb{H}$ that yield this distance.

6.3.4 Hyperbolic Flows

In this section we first introduce the notion of a hyperbolic set for a flow. We then show that for the geodesic flow in hyperbolic geometry any compact quotient is a hyperbolic set.

Let $\varphi_t \colon M \to M$ be a flow of a manifold M. We always assume that the map $(t, x) \mapsto \varphi_t(x)$ is of class C^1. We recall that a set $\Lambda \subset M$ is said to be Φ-invariant, where $\Phi = (\varphi_t)_{t \in \mathbb{R}}$, if $\varphi_t(\Lambda) = \Lambda$ for $t \in \mathbb{R}$.

Definition 6.8 A compact Φ-invariant set $\Lambda \subset M$ is said to be a *hyperbolic set* for Φ if there exist $\lambda \in (0, 1)$, $c > 0$, and decompositions

$$T_x M = E^s(x) \oplus E^0(x) \oplus E^u(x) \quad \text{for } x \in \Lambda$$

such that, for any $x \in \Lambda$:

1. $E^0(x)$ is the space of dimension 1 generated by the vector

$$X(x) = \frac{d}{dt} \varphi_t(x) \bigg|_{t=0} ;$$

2. for each $t \in \mathbb{R}$,

$$d_x \varphi_t E^s(x) = E^s(\varphi_t(x)) \quad \text{and} \quad d_x \varphi_t E^u(x) = E^u(\varphi_t(x)); \qquad (6.75)$$

3. for each $t > 0$,

$$\|d_x \varphi_t v\| \le c \lambda^t \|v\| \quad \text{for } v \in E^s(x)$$

and

$$\|d_x \varphi_{-t} v\| \le c \lambda^t \|v\| \quad \text{for } v \in E^u(x).$$

The linear spaces $E^s(x)$ and $E^u(x)$ are called, respectively, the *stable* and *unstable* *spaces* at the point x.

Now we consider the particular case of the geodesic flow in hyperbolic geometry. For a quotient $\mathbb{H}/G$ as in Sect. 6.3.2, the inner product $\langle \cdot, \cdot \rangle_{(z,v)}$ on each tangent space

$$T_{(z,v)} S(\mathbb{H}/G) = T_{(z,v)} S\mathbb{H}$$

is the one yielding the distance d on $S(\mathbb{H}/G)$ (see (6.74)).

Theorem 6.5 *For the compact surface $\mathbb{H}/G$ of genus $g \ge 2$ in Proposition 6.5, the unit tangent bundle $S(\mathbb{H}/G)$ is a hyperbolic set for the geodesic flow. Moreover, for each $(z, v) \in S(\mathbb{H}/G)$, there exist manifolds*

$$V^s(z, v), V^u(z, v) \subset S(\mathbb{H}/G)$$

Fig. 6.8 The sets $H_{z,v}^s$ and $H_{z,v}^u$

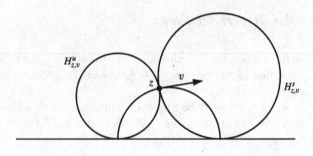

Fig. 6.9 The sets $V^s(z,v)$ and $V^u(z,v)$

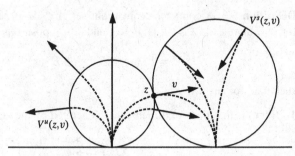

containing (z,v) that are tangent, respectively, to the stable and unstable spaces at (z,v) and which satisfy

$$\varphi_t\big(V^u(z,v)\big) = V^u(\varphi_t(z,v)) \quad and \quad \varphi_t\big(V^s(z,v)\big) = V^s(\varphi_t(z,v)) \qquad (6.76)$$

for any $(z,v) \in S\mathbb{H}$ and $t \in \mathbb{R}$.

Proof Given $(z,v) \in S\mathbb{H}$, let $\gamma(t)$ be the geodesic with $\gamma(0) = z$ and $\gamma'(0) = v$. Moreover, let $H_{z,v}^s$ and $H_{z,v}^u$ be the circles passing through z that are tangent to the real axis, respectively, at the points $\gamma(+\infty)$ and $\gamma(-\infty)$ (see Fig. 6.8). Now we consider the set

$$V^s(z,v) \subset S(\mathbb{H}/G)$$

formed by the vectors with norm 1 that are on $H_{z,v}^s$, are normal to $H_{z,v}^s$, and point in the same direction as v (see Fig. 6.9). Analogously, we consider the set

$$V^u(z,v) \subset S(\mathbb{H}/G)$$

formed by the vectors with norm 1 that are on $H_{z,v}^u$, are normal to $H_{z,v}^u$, and point in the same direction as v (see Fig. 6.9). One can verify that the sets $V^s(z,v)$ and $V^u(z,v)$ are manifolds of dimension 1. Moreover, property (6.76) holds. Indeed, the restriction of the geodesic flow to $\mathbb{H}$ maps circles tangent to the real line and horizontal lines to circles tangent to the real line or horizontal lines. On the other hand, by (6.67), the angles are preserved under the action of the geodesic flow and thus,

Fig. 6.10 The sets $H_{i,i}^s$ and $H_{i,i}^u$

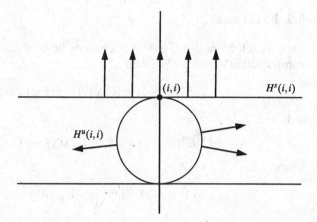

the image of a normal vector to $H_{z,v}^s$ or $H_{z,v}^u$ is still a normal vector, respectively, to the images of $H_{z,v}^s$ and $H_{z,v}^u$.

It remains to establish the third property in Definition 6.8 since then property (6.75) follows from (6.76). We first note that it is sufficient to consider the geodesic traversing the positive part of the imaginary axis, that is, the path $\gamma : \mathbb{R} \to \mathbb{H}$ given by $\gamma(t) = ie^t$. In this case, $H_{i,i}^u$ is the circle having as diameter the line segment between 0 and i, while $H_{i,i}^s$ is the horizontal straight line passing through i (see Fig. 6.10). For each $t > 0$, we have

$$d_{(i,i)}\varphi_t(1,0) = \lim_{h \to 0} \frac{\varphi_t((i,i) + h(1,0)) - \varphi_t(i,i)}{h}$$

$$= \lim_{h \to 0} \frac{(ie^t + h, ie^t) - (ie^t, ie^t)}{h} = (1 + i0, 0) \qquad (6.77)$$

and hence,

$$\|d_{(i,i)}\varphi_t(1,0)\|_{\varphi_t(i,i)} = \|1 + i0\|_{ie^t} = e^{-t}. \qquad (6.78)$$

Finally, we note that the Möbius transformation $T(z) = -1/z$ takes $H_{i,i}^s$ to $H_{i,i}^u$, while the derivative $T'(i) = -1$ reverses the direction of all vectors with norm 1. Together with (6.77) and (6.78), this implies that

$$\|d_{(i,i)}\varphi_{-t}v\| = e^{-t}\|v\|$$

for $v \in T_{(i,i)}V^u(i,i)$ and $t > 0$. Therefore, $S(\mathbb{H}/G)$ is a hyperbolic set for the geodesic flow.   □

The sets $V^s(z, v)$ and $V^u(z, v)$ are called, respectively, the *stable* and *unstable* *manifolds* at the point (z, v).

6.4 Exercises

Exercise 6.1 Show that if $0 \in \mathbb{R}^p$ is a hyperbolic fixed point for an invertible linear transformation $A \colon \mathbb{R}^p \to \mathbb{R}^p$, then

$$E^s(0) = \left\{ y \in \mathbb{R}^p \setminus \{0\} : \lambda(y) < 0 \right\} \cup \{0\}$$

and

$$E^u(0) = \left\{ y \in \mathbb{R}^p \setminus \{0\} : \lambda(y) > 0 \right\} \cup \{0\},$$

where

$$\lambda(y) = \limsup_{n \to \infty} \frac{1}{n} \log \left\| A^n y \right\|. \tag{6.79}$$

Exercise 6.2 Given a $p \times p$ matrix A, show that any two vectors $v_1, v_2 \in \mathbb{R}^p \setminus \{0\}$ with $\lambda(v_1) \neq \lambda(v_2)$ (see (6.79)) are linearly independent.

Exercise 6.3 Let f be a diffeomorphism. Show that Λ is a hyperbolic set for f if and only if Λ is a hyperbolic set for f^{-1}.

Exercise 6.4 Determine whether a stable manifold can contain two periodic points.

Exercise 6.5 Show that the Smale horseshoe is locally maximal.

Exercise 6.6 Show that the solenoid Λ in Exercise 5.18 is locally maximal.

Exercise 6.7 Let Λ be a hyperbolic set for a diffeomorphism $f \colon \mathbb{R}^p \to \mathbb{R}^p$ with $E^u(x) = \{0\}$ for each $x \in \Lambda$. Show that Λ is the union of a finite number of periodic points of f.

Exercise 6.8 Let x be a hyperbolic fixed point of a diffeomorphism f. Show that for each $n \in \mathbb{N}$, there exists an open neighborhood V of x such that any periodic point of f in V has period greater than n.

Exercise 6.9 Determine whether there exists a homeomorphism $f \colon \mathbb{T}^2 \to \mathbb{T}^2$ such that $f \circ T_A = T_B \circ f$, where

$$A = \begin{pmatrix} 2 & 1 \\ 1 & 1 \end{pmatrix} \quad \text{and} \quad B = \begin{pmatrix} 3 & 1 \\ 1 & 1 \end{pmatrix}.$$

Exercise 6.10 Construct a topological conjugacy $h \colon \mathbb{R}^2 \to \mathbb{R}^2$ between the flows determined by the differential equations $x' = 3x$ and $x' = 4x$, that is, construct a homeomorphism h such that

$$h \circ \varphi_t = \psi_t \circ h \quad \text{for } t \in \mathbb{R},$$

where φ_t and ψ_t are the flows determined by the two equations.

Exercise 6.11 Determine whether the topological conjugacy $h: \mathbb{R}^2 \to \mathbb{R}^2$ constructed in Exercise 6.10 can be a diffeomorphism.

Exercise 6.12 Determine whether, up to compactness, the set $\mathbb{R}^2$ satisfies the conditions in the notion of a hyperbolic set for the flow $\varphi_t: \mathbb{R}^2 \to \mathbb{R}^2$ defined by

$$\varphi_t(x, y) = \left(e^t x, e^{-t} y\right).$$

Exercise 6.13 Determine whether, up to compactness, the set $\mathbb{R}^3$ satisfies the conditions in the notion of a hyperbolic set for the flow $\varphi_t: \mathbb{R}^3 \to \mathbb{R}^3$ defined by

$$\varphi_t(s, x, y) = \left(s + t, e^t x, e^{-t} y\right).$$

Exercise 6.14 Determine whether there exists a topological conjugacy $h: \mathbb{R}^2 \to \mathbb{R}^2$ between the flows defined by the equations

$$\begin{cases} x' = y, \\ y' = -x \end{cases} \quad \text{and} \quad \begin{cases} x' = y, \\ y' = -2x. \end{cases}$$

Exercise 6.15 Determine whether there exists a topological conjugacy $h: \mathbb{R}^2 \to \mathbb{R}^2$ between the flows determined by the equations

$$\begin{cases} x' = y, \\ y' = x \end{cases} \quad \text{and} \quad \begin{cases} x' = y, \\ y' = 3x. \end{cases}$$

Exercise 6.16 Show that Möbius transformations map straight lines and circles to straight lines or circles.

Exercise 6.17 Show that if Λ is a hyperbolic set for a flow $\Phi = (\varphi_t)_{t \in \mathbb{R}}$, then Λ is not a hyperbolic set for the diffeomorphism φ_T, for any $T \in \mathbb{R}$.

Exercise 6.18 Show that if Λ is a hyperbolic set for a flow Φ, then the stable and unstable subspaces $E^s(x)$ and $E^u(x)$ vary continuously with $x \in \Lambda$.

Exercise 6.19 Give an example of a hyperbolic set Λ for a diffeomorphism f such that $h(f|\Lambda) = 0$.

Exercise 6.20 Show that any Anosov diffeomorphism has positive topological entropy.

Chapter 7
Symbolic Dynamics

This chapter is an introduction to symbolic dynamics, with emphasis on its relations to hyperbolic dynamics. In particular, it is sometimes easier to solve certain problems of hyperbolic dynamics, such as those concerning periodic points, after associating a symbolic dynamics (also called a coding) to a hyperbolic set. After introducing some basic notions of symbolic dynamics, we illustrate with several examples how one can associate naturally a coding to several dynamical systems considered in the former chapters. These include expanding maps, quadratic maps and the Smale horseshoe. We also consider topological Markov chains, and we study their periodic points, topological entropy, and recurrence properties. Finally, we consider briefly the notion of the zeta function of a dynamical system.

7.1 Basic Notions

In this section we introduce some basic notions of symbolic dynamics. We also compute the topological entropy of the shift map.

7.1.1 Shift Map

Given an integer $k > 1$, consider the set $\Sigma_k^+ = \{1, \ldots, k\}^{\mathbb{N}}$ of sequences

$$\omega = (i_1(\omega)i_2(\omega)\cdots),$$

where $i_n(\omega) \in \{1, \ldots, k\}$ for each $n \in \mathbb{N}$.

Definition 7.1 The *shift map* $\sigma : \Sigma_k^+ \to \Sigma_k^+$ is defined by

$$\sigma(\omega) = (i_2(\omega)i_3(\omega)\cdots). \tag{7.1}$$

L. Barreira, C. Valls, *Dynamical Systems*, Universitext, DOI 10.1007/978-1-4471-4835-7_7,   153
© Springer-Verlag London 2013

Clearly, the map σ is not invertible.

Example 7.1 Given $m \in \mathbb{N}$, we compute the number of m-periodic points of σ. These are the sequences $\omega \in \Sigma_k^+$ such that $\sigma^m(\omega) = \omega$. It follows from (7.1) that ω is m-periodic if and only if

$$i_{n+m}(\omega) = i_n(\omega) \quad \text{for } n \in \mathbb{N}, \tag{7.2}$$

or equivalently, the first m elements of ω are repeated indefinitely. Thus, in order to specify an m-periodic point it is sufficient to specify its first m elements. On the other hand, given integers $j_1, \ldots, j_m \in \{1, \ldots, k\}$, the sequence $\omega \in \Sigma_k^+$ with

$$i_n(\omega) = j_n \quad \text{for } n = 1, \ldots, m$$

that satisfies (7.2) is an m-periodic point. Thus, the number of m-periodic points of σ is equal to

$$\text{card}\big(\{1, \ldots, k\}^m\big) = k^m.$$

Now we introduce a distance and thus also a topology on Σ_k^+. Given $\beta > 1$, for each $\omega, \omega' \in \Sigma_k^+$, let

$$d(\omega, \omega') = \begin{cases} \beta^{-n} & \text{if } \omega \neq \omega', \\ 0 & \text{if } \omega = \omega', \end{cases} \tag{7.3}$$

where $n = n(\omega, \omega') \in \mathbb{N}$ is the smallest positive integer such that $i_n(\omega) \neq i_n(\omega')$.

Proposition 7.1 *For each $\beta > 1$, the following properties hold:*

1. *d is a distance on Σ_k^+;*
2. *(Σ_k^+, d) is a compact metric space;*
3. *the shift map $\sigma: \Sigma_k^+ \to \Sigma_k^+$ is continuous.*

Proof It follows from (7.3) that

$$d(\omega', \omega) = d(\omega, \omega')$$

and that $d(\omega, \omega') = 0$ if and only if $\omega = \omega'$. Moreover, given $\omega, \omega', \omega'' \in \Sigma_k^+$, we have

$$d(\omega, \omega'') = \beta^{-n_1}, \qquad d(\omega, \omega') = \beta^{-n_2}, \qquad d(\omega', \omega'') = \beta^{-n_3},$$

where n_1, n_2 and n_3 are, respectively, the smallest positive integers such that

$$i_{n_1}(\omega) \neq i_{n_1}(\omega''), \qquad i_{n_2}(\omega) \neq i_{n_2}(\omega'), \qquad i_{n_3}(\omega') \neq i_{n_3}(\omega''). \tag{7.4}$$

We note that if $n_2 > n_1$ and $n_3 > n_1$, then $i_{n_1}(\omega) = i_{n_1}(\omega') = i_{n_1}(\omega'')$, which contradicts (7.4). Hence, $n_2 \leq n_1$ or $n_3 \leq n_1$ and thus,

$$\beta^{-n_1} \leq \beta^{-n_2} \quad \text{or} \quad \beta^{-n_1} \leq \beta^{-n_3}.$$

This establishes the triangle inequality.

In order to show that Σ_k^+ is compact, we first note that the sets

$$C_{j_1 \cdots j_m} = \{\omega \in \Sigma_k^+ : i_n(\omega) = j_n \text{ for } n = 1, \ldots, m\}, \qquad (7.5)$$

with $j_1, \ldots, j_m \in \{1, \ldots, k\}$, are exactly the d-open balls. Equipping $\{1, \ldots, k\}$ with the discrete topology (in which all subsets of $\{1, \ldots, k\}$ are open), the product topology on $\Sigma_k^+ = \{1, \ldots, k\}^{\mathbb{N}}$ coincides with the topology generated by the open balls $C_{j_1 \cdots j_m}$ in (7.5). In other words, it coincides with the topology induced by the distance d. It follows from Tychonoff's theorem[1] that (Σ_k^+, d) is a compact topological space (it is the product of compact topological spaces, with the product topology).

For the last property, we note that if $d(\omega, \omega') = \beta^{-n}$, then

$$d(\sigma(\omega), \sigma(\omega')) \leq \beta^{-(n-1)} = \beta d(\omega, \omega')$$

and the shift map is continuous. □

It also follows from the proof of Proposition 7.1 that

$$d(\omega, \omega'') \leq \max\{d(\omega, \omega'), d(\omega', \omega'')\}.$$

7.1.2 Topological Entropy

By Proposition 7.1, $\sigma : \Sigma_k^+ \to \Sigma_k^+$ is a continuous map of a compact metric space. Hence, its topological entropy is well defined (see Definition 3.8).

Proposition 7.2 We have $h(\sigma|\Sigma_k^+) = \log k$.

Proof Given $m, p \in \mathbb{N}$ and $\omega, \omega' \in \Sigma_k^+$, we have

$$d_m(\omega, \omega') = \max\{d(\sigma^j(\omega), \sigma^j(\omega')) : j = 0, \ldots, m - 1\}.$$

Clearly, $d(\sigma^j(\omega), \sigma^j(\omega')) \geq \beta^{-p}$ if and only if

$$n = n(\omega, \omega') \in \{1 + j, \ldots, p + j\}$$

---

[1] **Theorem** (See for example [43]) Any product of compact topological spaces equipped with the product topology is a compact topological space.

and thus,

$$d_m(\omega, \omega') \geq \beta^{-p} \quad \text{if and only if} \quad n \leq p + m - 1. \tag{7.6}$$

This implies that

$$N(m, \beta^{-p}) \leq k^{p+m-1} \tag{7.7}$$

since the right-hand side is exactly the largest number of distinct sequences in Σ_k^+ that differ in some of their first $p + m - 1$ elements.

Now we note that the number of $(p + m - 1)$-periodic points of σ is k^{p+m-1}. If ω and ω' are two of these points, then

$$d_m(\omega, \omega') = \max\{d(\sigma^j(\omega), \sigma^j(\omega')) : j = 0, \ldots, m - 1\} \geq \beta^{-p} \tag{7.8}$$

since

$$n(\omega, \omega') \in \{1, \ldots, p + m - 1\}.$$

Hence, $N(m, \beta^{-p}) \geq k^{p+m-1}$ and it follows from (7.7) that

$$N(m, \beta^{-p}) = k^{p+m-1}.$$

Finally,

$$h(\sigma|\Sigma_k^+) = \lim_{p \to \infty} \lim_{m \to \infty} \frac{1}{m} \log N(m, \beta^{-p})$$

$$= \lim_{p \to \infty} \lim_{m \to \infty} \frac{p + m - 1}{m} \log k = \log k,$$

which yields the desired result.   □

7.1.3 Two-Sided Sequences

One can consider in an analogous manner the case of two-sided sequences. Namely, given an integer $k > 1$, consider the set $\Sigma_k = \{1, \ldots, k\}^{\mathbb{Z}}$ of sequences

$$\omega = (\cdots i_{-1}(\omega) i_0(\omega) i_1(\omega) \cdots).$$

Definition 7.2 The *shift map* $\sigma : \Sigma_k \to \Sigma_k$ is defined by $\sigma(\omega) = \omega'$, where

$$i_n(\omega') = i_{n+1}(\omega) \quad \text{for } n \in \mathbb{Z}.$$

We note that the shift map on Σ_k is invertible.

Example 7.2 In an analogous manner to that in Example 7.1, given $m \in \mathbb{N}$, a point $\omega \in \Sigma_k$ is m-periodic if and only if

$$i_{n+m}(\omega) = i_n(\omega) \quad \text{for } n \in \mathbb{Z}. \tag{7.9}$$

Hence, in order to specify an m-periodic point $\omega \in \Sigma_k$ it is sufficient to specify the elements $i_1(\omega), \ldots, i_m(\omega)$. On the other hand, given integers $j_1, \ldots, j_m \in \{1, \ldots, k\}$, the sequence $\omega \in \Sigma_k$ with

$$i_n(\omega) = j_n \quad \text{for } n = 1, \ldots, m$$

that satisfies (7.9) is an m-periodic point. This implies that the number of m-periodic points of $\sigma | \Sigma_k$ is equal to k^m.

Now we introduce a distance and thus also a topology on Σ_k. Given $\beta > 1$, for each $\omega, \omega' \in \Sigma_k$, let

$$d(\omega, \omega') = \begin{cases} \beta^{-n} & \text{if } \omega \neq \omega', \\ 0 & \text{if } \omega = \omega', \end{cases}$$

where $n = n(\omega, \omega') \in \mathbb{N}$ is the smallest integer such that

$$i_n(\omega) \neq i_n(\omega') \quad \text{or} \quad i_{-n}(\omega) \neq i_{-n}(\omega').$$

One can verify that d is a distance on Σ_k.

7.2 Examples of Codings

In this section we illustrate how one can naturally associate a symbolic dynamics (that is, a shift map on some space Σ_k^+ or Σ_k), also known as a *coding*, to several dynamical systems introduced in the former chapters.

7.2.1 Expanding Maps

We first consider the expanding maps and their topological entropy.

Example 7.3 Consider the expanding map $E_2 \colon S^1 \to S^1$. As we observed in Example 3.4, writing $x = 0.x_1 x_2 \cdots \in S^1$ in base-2 (with $x_n \in \{0, 1\}$ for each n), we have

$$E_2(0.x_1 x_2 \cdots) = 0.x_2 x_3 \cdots.$$

This is precisely the behavior observed in (7.1) and thus, it is natural to expect that there exists some relation between E_2 and $\sigma | \Sigma_2^+$.

We define a function $H: \Sigma_2^+ \to S^1$ by

$$H(i_1 i_2 \cdots) = \sum_{n=1}^{\infty} (i_n - 1)2^{-n} = 0.(i_1 - 1)(i_2 - 1) \cdots . \qquad (7.10)$$

Then

$$\begin{aligned}
(H \circ \sigma)(i_1 i_2 \cdots) &= H(i_2 i_3 \cdots) = \sum_{n=1}^{\infty} (i_{n+1} - 1)2^{-n} \\
&= 0.(i_2 - 1)(i_3 - 1) \cdots \\
&= E_2 \big(0.(i_1 - 1)(i_2 - 1) \cdots \big) \\
&= (E_2 \circ H)(i_1 i_2 \cdots),
\end{aligned}$$

that is,

$$H \circ \sigma = E_2 \circ H \quad \text{in } \Sigma_2^+. \qquad (7.11)$$

We note that the map H is not one-to-one since

$$H(i_1 \cdots i_n 211 \cdots) = H(i_1 \cdots i_n 122 \cdots)$$

for any $i_1, \ldots, i_n \in \{1, 2\}$. However, if $B \subset \Sigma_2^+$ is the subset of all sequences with infinitely many consecutive 2's, then the map

$$H|(\Sigma_2^+ \setminus B): \Sigma_2^+ \setminus B \to S^1 \qquad (7.12)$$

is bijective.

Example 7.4 Now we use the former example to find the number of m-periodic points of the expanding map E_2. By Example 7.1, the number of m-periodic points of the shift map $\sigma | \Sigma_2^+$ is 2^m. Only one of them belongs to B (see (7.12)), namely the constant sequence $(22 \cdots)$. Thus, the number of m-periodic points of $\sigma | (\Sigma_2^+ \setminus B)$ is $2^m - 1$. We note that the set $\Sigma_2^+ \setminus B$ is forward σ-invariant and hence, the orbits of these points are in fact in $\Sigma_2^+ \setminus B$.

On the other hand, it follows from (7.11) that

$$H \circ \sigma^m = E_2^m \circ H \quad \text{in } \Sigma_2^+, \qquad (7.13)$$

for each $m \in \mathbb{N}$. Now take $\omega \in \Sigma_2^+ \setminus B$ and $m \in \mathbb{N}$. Since the set $\Sigma_2^+ \setminus B$ is forward σ-invariant, we have $\sigma^m(\omega) \in \Sigma_2^+ \setminus B$. Moreover, since the function $H|(\Sigma_2^+ \setminus B)$ is bijective, it follows from (7.13) that $\sigma^m(\omega) = \omega$ if and only if

$$H(\omega) = H\big(\sigma^m(\omega)\big) = E_2^m(H(\omega)).$$

Thus, $\omega \in \Sigma_2^+ \setminus B$ is an m-periodic point of σ if and only if $H(\omega)$ is an m-periodic point of E_2. This implies that the number of m-periodic points of the expanding map E_2 is $2^m - 1$ (as we already saw in Sect. 2.2.2), namely

$$x_{i_1 \cdots i_m} = H(i_1 \cdots i_m i_1 \cdots i_m \cdots) \in S^1$$

for $(i_1, \ldots, i_m) \in \{1, \ldots, k\}^m \setminus \{(2, \ldots, 2)\}$. It follows from (7.10) that

$$x_{i_1 \cdots i_m} = \sum_{n=1}^{m} (i_n - 1)2^{-n}\left(1 + 2^{-m} + 2^{-2m} + \cdots\right)$$

$$= \frac{1}{1 - 2^{-m}} \sum_{n=1}^{m} (i_n - 1)2^{-n}$$

$$= \frac{1}{2^m - 1} \sum_{n=1}^{m} (i_n - 1)2^{m-n}.$$

The sum $\sum_{n=1}^{m} (i_n - 1)2^{m-n}$ takes the values $0, 1, \ldots, 2^m - 1$ since $(i_1, \ldots, i_m) \neq (2, \ldots, 2)$. Hence, we recover the periodic points already obtained in (2.2).

The following example also illustrates how a coding can be used to compute the topological entropy.

Example 7.5 Consider the restriction $E_4|A \colon A \to A$ of the map E_4, where A is the compact forward E_4-invariant set in (2.11). Writing $x = 0.x_1 x_2 \cdots \in S^1$ in base-4, with $x_n \in \{0, 1, 2, 3\}$ for each $n \in \mathbb{N}$, we have

$$E_4(0.x_1 x_2 \cdots) = 0.x_2 x_3 \cdots.$$

Now we define a function $H \colon \Sigma_2^+ \to S^1$ by

$$H(i_1 i_2 \cdots) = \sum_{n=1}^{\infty} 2(i_1 - 1)4^{-n} = 0.j_1 j_2 \cdots,$$

also in base-4, where

$$j_n = 2(i_n - 1) \in \{0, 2\} \quad \text{for } n \in \mathbb{N}. \tag{7.14}$$

We have

$$(H \circ \sigma)(i_1 i_2 \cdots) = H(i_2 i_3 \cdots) = \sum_{n=1}^{\infty} 2(i_{n+1} - 1)4^{-n}$$

and

$$(E_4 \circ H)(i_1 i_2 \cdots) = E_4(0.j_1 j_2 \cdots) = 0.j_2 j_3 \cdots.$$

Hence, it follows from (7.14) that

$$H \circ \sigma = E_4 \circ H \quad \text{in } \Sigma_2^+.$$

We note that the map H is one-to-one, unlike in Example 7.3. It is also a homeomorphism onto its image $H(\Sigma_2^+) = A$. Indeed, given $\omega, \omega' \in \Sigma_2^+$ with $\omega \neq \omega'$, we have

$$d_{S^1}\big(H(\omega), H(\omega')\big) \leq \sum_{m=n}^{\infty} 2 \cdot 4^{-m} = \frac{8 \cdot 4^{-n}}{3}$$

$$= \frac{8}{3}\big(\beta^{-n}\big)^{\log 4 / \log \beta}$$

$$= \frac{8}{3}d(\omega, \omega')^{\log 4 / \log \beta},$$

where $n = n(\omega, \omega') \in \mathbb{N}$ is the smallest integer such that $i_n(\omega) \neq i_n(\omega')$ and where d_{S^1} is the distance on S^1. On the other hand, given

$$x = 0.j_1 j_2 \cdots, \quad x' = 0.j_1' j_2' \cdots \in A,$$

or equivalently, $(j_1 j_2 \cdots), (j_1' j_2' \cdots) \in \{0, 2\}^{\mathbb{N}}$, we have

$$d\big(H^{-1}(x), H^{-1}(x')\big) = d\left(\sum_{n=1}^{\infty} 2(i_n - 1)4^{-n}, \sum_{n=1}^{\infty} 2(i_n' - 1)4^{-n}\right),$$

with

$$j_n = 2(i_n - 1) \quad \text{and} \quad j_n' = 2(i_n' - 1)$$

for $n \in \mathbb{N}$. Now take $x \neq x'$ such that

$$d_{S^1}(x, x') = |x - x'|.$$

If $n \in \mathbb{N}$ is the smallest integer such that $j_n \neq j_n'$ or, equivalently, $i_n \neq i_n'$, then

$$d_{S^1}(x, x') \geq 2 \cdot 4^{-n} - \sum_{m=n+1}^{\infty} 2 \cdot 4^{-m} = \frac{1}{3}4^{-n+1}$$

and

$$d\big(H^{-1}(x), H^{-1}(x')\big) = \beta^{-n} = 4^{-n \log \beta / \log 4}$$

$$= \left(\frac{3}{4} \cdot \frac{1}{3}4^{-n+1}\right)^{\log \beta / \log 4}$$

$$\leq \left(\frac{3}{4}d_{S^1}(x, x')\right)^{\log \beta / \log 4}.$$

This shows that $H: \Sigma_2^+ \to A$ is a homeomorphism. Finally, it follows from Theorem 3.3 together with Proposition 7.2 that

$$h(E_4|A) = h(\sigma|\Sigma_2^+) = \log 2,$$

as we already obtained in Example 3.18.

7.2.2 Quadratic Maps

In this section we consider a class of quadratic maps.

Example 7.6 Given $a > 4$, let $f: [0, 1] \to \mathbb{R}$ be the quadratic map

$$f(x) = ax(1 - x)$$

and let $X \subset [0, 1]$ be the forward f-invariant set in (3.27). We also consider the restriction $f|X: X \to X$. Now we define a function $H: \Sigma_2^+ \to X$ by

$$H(i_1 i_2 \cdots) = \bigcap_{n=1}^{\infty} f^{-n+1} I_{i_n}, \qquad (7.15)$$

where

$$I_1 = \left[0, (1 - \sqrt{1 - 4/a})/2\right] \quad \text{and} \quad I_2 = \left[(1 + \sqrt{1 - 4/a})/2, 1\right].$$

We show that for any sufficiently large a the map H is well defined, that is, the intersection in (7.15) contains exactly one point for each sequence $(i_1 i_2 \cdots) \in \Sigma_2^+$. Given $a > 2 + \sqrt{5}$, we have

$$|f'(x)| = a|1 - 2x| \geq \lambda > 1 \qquad (7.16)$$

for $x \in I_1 \cup I_2$, where $\lambda = a\sqrt{1 - 4/a}$. Hence, each interval

$$I_{i_1 \cdots i_m} = \bigcap_{n=1}^{m} f^{-n+1} I_{i_n}$$

has length at most $\lambda^{-(m-1)}$ and thus, each intersection in (7.15) contains exactly one point. Since $f^{-1}[0, 1] = I_1 \cup I_2$, it follows from (3.27) that

$$X = \bigcap_{n=0}^{\infty} f^{-n}(I_1 \cup I_2) = \bigcup_{(i_1 i_2 \cdots) \in \Sigma_2^+} H(i_1 i_2 \cdots),$$

and the map H is onto. It is also invertible, with its inverse given by $H^{-1}(x) = (i_1 i_2 \cdots)$, where $i_n = j$ when $f^{n-1}(x) \in I_j$, for each $n \in \mathbb{N}$.

We also show that H is a homeomorphism. Given distinct points $\omega, \omega' \in \Sigma_2^+$ with $n = n(\omega, \omega') > 1$, we have

$$\left| H(\omega) - H(\omega') \right| = a_{i_1 \cdots i_{n-1}},$$

where $a_{i_1 \cdots i_{n-1}}$ is the length of the interval $I_{i_1 \cdots i_{n-1}}$. It follows from (7.16) that

$$\left| H(\omega) - H(\omega') \right| \le \lambda^{-(n-2)} \to 0$$

when $n \to \infty$. This shows that the map H is continuous. On the other hand, given distinct points $x, x' \in X$, there exists an $n \in \mathbb{N}$ such that

$$I_{i_1 \cdots i_{n-1}} = I_{i'_1 \cdots i'_{n-1}} \quad \text{and} \quad I_{i_1 \cdots i_n} \cap I_{i'_1 \cdots i'_n} = \varnothing, \tag{7.17}$$

where

$$H^{-1}(x) = (i_1 i_2 \cdots) \quad \text{and} \quad H^{-1}(x') = (i'_1 i'_2 \cdots).$$

Then

$$d\left(H^{-1}(x), H^{-1}(x') \right) = \beta^{-n} \to 0$$

when $n \to \infty$. It follows from (7.17) that $|x - x'| \ge \lambda^{-(n-1)}$ and thus, if $x' \to x$, then $n \to \infty$. This shows that the map H^{-1} is continuous.

Since $H \colon \Sigma_2^+ \to X$ is a homeomorphism, it follows from Theorem 3.3 together with Proposition 7.2 that

$$h(f|X) = h(\sigma|\Sigma_2^+) = \log 2.$$

7.2.3 The Smale Horseshoe

We also associate a symbolic dynamics to the Smale horseshoe.

Example 7.7 Let $\Lambda \subset [0, 1]^2$ be the Smale horseshoe constructed in Sect. 5.2.2 from a diffeomorphism f defined in an open neighborhood of the square $[0, 1]^2$. We continue to consider the vertical strips V_1 and V_2 in (5.5) and we define a function $H \colon \Sigma_2 \to \Lambda$ by

$$H(\cdots i_{-1} i_0 i_1 \cdots) = \bigcap_{n \in \mathbb{Z}} f^{-n} V_{i_n}. \tag{7.18}$$

In order to verify that H is well defined, consider the sets

$$R_n(\omega) = \bigcap_{k=-n}^{n} f^{-k} V_{i_k},$$

where $\omega = (\cdots i_{-1} i_0 i_1 \cdots)$. Each $R_n(\omega)$ is contained in a square of size a^n and thus, diam $R_n(\omega) \to 0$ when $n \to \infty$. This implies that each intersection

$$\bigcap_{n \in \mathbb{Z}} f^{-n} V_{i_n} = \bigcap_{n \in \mathbb{Z}} R_n(\omega)$$

has at most one point. On the other hand, since $R_n(\omega)$ is a decreasing sequence of nonempty closed sets, the intersection $\bigcap_{n \in \mathbb{N}} R_n(\omega)$ has at least one point. This shows that card $H(\omega) = 1$ for each $\omega \in \Sigma_2$ and the function H is well defined.

Moreover, it follows from the construction of the Smale horseshoe that

$$\Lambda = \bigcap_{n \in \mathbb{Z}} f^{-n}(V_1 \cup V_2)$$

$$= \bigcup_{\omega \in \Sigma_2} \bigcap_{n \in \mathbb{Z}} f^{-n} V_{i_n} = \bigcup_{\omega \in \Sigma_2} H(\omega)$$

and thus, the map H is onto. In order to show that it is one-to-one, let us take sequences $\omega, \omega' \in \Sigma_2$ with $\omega \neq \omega'$. Then there exists an $m \in \mathbb{Z}$ such that $i_m(\omega) \neq i_m(\omega')$ and thus also

$$V_{i_m(\omega)} \cap V_{i_m(\omega')} = \varnothing.$$

Hence,

$$H(\omega) \cap H(\omega') = \left(\bigcap_{n \in \mathbb{Z}} f^{-n} V_{i_n(\omega)} \right) \cap \left(\bigcap_{n \in \mathbb{Z}} f^{-n} V_{i_n(\omega')} \right) = \varnothing.$$

This shows that $H(\omega) \neq H(\omega')$ and the map H is one-to-one.

We also have

$$H(\sigma(\omega)) = \bigcap_{n \in \mathbb{Z}} f^{-n} V_{i_{n+1}(\omega)}$$

$$= \bigcap_{n \in \mathbb{Z}} f^{1-n} V_{i_n(\omega)} = f(H(\omega)),$$

that is,

$$H \circ \sigma = f \circ H \quad \text{in } \Sigma_2.$$

Given $m \in \mathbb{N}$ and $\omega \in \Sigma_2$, we obtain

$$H(\sigma^m(\omega)) = f^m(H(\omega)).$$

This implies that ω is an m-periodic point of σ if and only if $H(\omega)$ is an m-periodic point of $f|\Lambda$. Moreover, ω is a periodic point of σ with period m if and only if $H(\omega)$ is a periodic point of $f|\Lambda$ with period m. In particular, it follows from Example 7.2 that the number of m-periodic points of $f|\Lambda$ is 2^m.

7.3 Topological Markov Chains

In this section we consider a class of subsets of Σ_k^+ that are forward σ-invariant. They give rise to topological Markov chains.

7.3.1 Basic Notions

Given an integer $k > 1$, let $A = (a_{ij})$ be a $k \times k$ matrix with entries $a_{ij} \in \{0, 1\}$ for each i and j. We consider the subset of Σ_k^+ defined by

$$\Sigma_A^+ = \left\{ \omega \in \Sigma_k^+ : a_{i_n(\omega)i_{n+1}(\omega)} = 1 \text{ for } n \in \mathbb{N} \right\}. \tag{7.19}$$

Clearly, $\sigma(\Sigma_A^+) \subset \Sigma_A^+$. This allows one to introduce the following notion.

Definition 7.3 The restriction $\sigma|\Sigma_A^+ \colon \Sigma_A^+ \to \Sigma_A^+$ is called the *topological Markov chain* with *transition matrix A*.

It is also common to use the alternative expression *(sub)shift of finite type* to refer to a topological Markov chain.

Example 7.8 Let A be the $k \times k$ matrix with all entries equal to 1. In this case, it follows from (7.19) that $\Sigma_A^+ = \Sigma_k^+$ and thus, the topological Markov chain $\sigma|\Sigma_A^+$ coincides with the shift map $\sigma \colon \Sigma_k^+ \to \Sigma_k^+$.

Example 7.9 For the matrix

$$A = \begin{pmatrix} 0 & 1 \\ 1 & 1 \end{pmatrix}, \tag{7.20}$$

we have

$$\Sigma_A^+ = \left\{ \omega \in \Sigma_2^+ : a_{i_n(\omega)i_{n+1}(\omega)} = 1 \text{ for } n \in \mathbb{N} \right\}$$
$$= \left\{ \omega \in \Sigma_2^+ : \left(i_n(\omega), i_{n+1}(\omega) \right) \neq (1, 1) \text{ for } n \in \mathbb{N} \right\}.$$

In other words, Σ_A^+ is the subset of all sequences in Σ_2^+ in which the symbol 1 is always isolated (whenever it occurs).

One can consider in an analogous manner the case of two-sided sequences. Namely, given an integer $k > 1$, let $A = (a_{ij})$ be a $k \times k$ matrix with entries $a_{ij} \in \{0, 1\}$ for each i and j. We consider the subset of Σ_k defined by

$$\Sigma_A = \left\{ \omega \in \Sigma_k : a_{i_n(\omega)i_{n+1}(\omega)} = 1 \text{ for } n \in \mathbb{Z} \right\}.$$

We have $\sigma(\Sigma_A) = \Sigma_A$ and thus, one can introduce the following notion.

Definition 7.4 The restriction $\sigma|\Sigma_A : \Sigma_A \to \Sigma_A$ is called the *(two-sided) topological Markov chain* with *transition matrix A*.

We also give some examples.

Example 7.10 For the matrix

$$A = \begin{pmatrix} 0 & 1 \\ 1 & 0 \end{pmatrix},$$

we have

$$\Sigma_A = \left\{\omega \in \Sigma_2 : a_{i_n(\omega)i_{n+1}(\omega)} = 1 \text{ for } n \in \mathbb{Z}\right\}$$
$$= \left\{\omega \in \Sigma_2 : i_n(\omega) \neq i_{n+1}(\omega) \text{ for } n \in \mathbb{Z}\right\}.$$

Hence, the set Σ_A has exactly two sequences, namely

$$\omega_1 = (\cdots i_0 \cdots) \quad \text{and} \quad \omega_2 = (\cdots j_0 \cdots),$$

where

$$i_n = \begin{cases} 1 & \text{if } n \text{ is even,} \\ 2 & \text{if } n \text{ is odd} \end{cases} \quad \text{and} \quad j_n = \begin{cases} 2 & \text{if } n \text{ is even,} \\ 1 & \text{if } n \text{ is odd.} \end{cases}$$

We note that $\sigma(\omega_1) = \omega_2$ and $\sigma(\omega_2) = \omega_1$. Thus, $\Sigma_A = \{\omega_1, \omega_2\}$ is a periodic orbit with period 2.

Example 7.11 Let $\Sigma \subset \Sigma_2$ be the subset of all sequences in Σ_2 in which the symbol 1 occurs finitely many times and always in pairs (when it occurs). Clearly, $\sigma(\Sigma) = \Sigma$ and one can consider the restriction $\sigma|\Sigma : \Sigma \to \Sigma$. Now we show that $\sigma|\Sigma$ is not a topological Markov chain. Consider the sequence $\omega = (\cdots i_0 \cdots)$ with $i_0 = i_1 = 1$ and $i_j = 2$ for $j \notin \{0, 1\}$. We note that $\omega \in \Sigma$. Thus, if $\sigma|\Sigma$ was a topological Markov chain, then we would have $\Sigma = \Sigma_2$. Indeed, the sequence ω contains the transitions $1 \to 1$, $1 \to 2$, $2 \to 1$ and $2 \to 2$. Since $\Sigma \neq \Sigma_2$, we conclude that $\sigma|\Sigma$ is not a topological Markov chain.

7.3.2 Periodic Points

In this section we compute the number of m-periodic points of an arbitrary topological Markov chain. We start with an example.

Example 7.12 Let $\sigma|\Sigma_A^+$ be the topological Markov chain with the transition matrix A in (7.20). We compute explicitly the number of m-periodic points for $m = 1$ and $m = 2$ (Example 7.13 considers the general case):

1. For $m = 1$, the sequence $(22\cdots)$ is the only fixed point of $\sigma|\Sigma_A^+$ since $a_{11} = 0$.

2. Now let $m = 2$. We note that a point $\omega \in \Sigma_A^+$ is m-periodic if and only if property (7.2) holds. Hence, we have to find the number of sequences in Σ_A^+ with this property, which coincides with the number of vectors $(i, j) \in \{1, 2\}^2$ such that the transitions

$$i \to j \to i$$

are allowed. This condition is equivalent to $a_{ij} = a_{ji} = 1$ and thus, the number of 2-periodic points of $\sigma | \Sigma_A^+$ is equal to

$$\sum_{i=1}^{2} \sum_{j=1}^{2} a_{ij} a_{ji} = \sum_{i=1}^{2} (A^2)_{ii} = \mathrm{tr}(A^2),$$

where $(A^2)_{ii}$ is the entry (i, i) of the matrix A^2.

Now we consider arbitrary matrices.

Proposition 7.3 *For each $m \in \mathbb{N}$, the number of m-periodic points of the topological Markov chain $\sigma | \Sigma_A^+$ is equal to* $\mathrm{tr}(A^m)$.

Proof We proceed in an analogous manner to that in Example 7.12. Since the point $\omega \in \Sigma_A^+$ is m-periodic if and only if property (7.2) holds, we have to find the number of sequences in Σ_A^+ with this property. This coincides with the number of vectors $(i_1, \ldots, i_m) \in \{1, \ldots, k\}^m$ such that the transitions

$$i_1 \to i_2 \to \cdots \to i_m \to i_1$$

are allowed. This condition is equivalent to

$$a_{i_1 i_2} = a_{i_2 i_3} = \cdots = a_{i_{m-1} i_m} = a_{i_m i_1} = 1$$

and thus, the number of m-periodic points of $\sigma | \Sigma_A^+$ is equal to

$$\sum_{(i_1, \ldots, i_m) \in \{1, \ldots, k\}^m} a_{i_1 i_2} a_{i_2 i_3} \cdots a_{i_m i_1} = \sum_{i_1 \in \{1, \ldots, k\}} (A^m)_{i_1 i_1} = \mathrm{tr}(A^m).$$

This yields the desired result. $\square$

Example 7.13 Let A be the matrix in (7.20). By Proposition 7.3, for each $m \in \mathbb{N}$, the number of m-periodic points of $\sigma | \Sigma_A^+$ is equal to $\mathrm{tr}(A^m)$. On the other hand, we have

$$A = \begin{pmatrix} 0 & 1 \\ 1 & 1 \end{pmatrix} = S \begin{pmatrix} (1 + \sqrt{5})/2 & 0 \\ 0 & (1 - \sqrt{5})/2 \end{pmatrix} S^{-1},$$

where

$$S = \begin{pmatrix} (-1 + \sqrt{5})/2 & (-1 - \sqrt{5})/2 \\ 1 & 1 \end{pmatrix}$$

and thus,

$$A^m = S \begin{pmatrix} (1+\sqrt{5})/2 & 0 \\ 0 & (1-\sqrt{5})/2 \end{pmatrix}^m S^{-1}.$$

Hence,

$$\operatorname{tr}(A^m) = \left(\frac{1+\sqrt{5}}{2}\right)^m + \left(\frac{1-\sqrt{5}}{2}\right)^m$$

and this is the number of m-periodic points of $\sigma|\Sigma_A^+$ (which must be an integer). For example,

$$\operatorname{tr}(A^3) = 4, \quad \operatorname{tr}(A^7) = 29 \quad \text{and} \quad \operatorname{tr}(A^{11}) = 199.$$

7.3.3 Topological Entropy

Now we compute the topological entropy of an arbitrary topological Markov chain.

Theorem 7.1 *We have $h(\sigma|\Sigma_A^+) = \log \rho(A)$, where $\rho(A)$ is the spectral radius of A.*

Proof We proceed in an analogous manner to that in the proof of Proposition 7.2. We first observe that since the map $\sigma|\Sigma_k^+$ is expansive, the same happens to the topological Markov chain $\sigma|\Sigma_A^+$ and thus, one can apply Theorem 3.5.

Given $m, p \in \mathbb{N}$ and $\omega, \omega' \in \Sigma_k^+$, by (7.6), we have

$$d_m(\omega, \omega') \geq \beta^{-p} \quad \text{if and only if} \quad n = n(\omega, \omega') \leq p + m - 1.$$

Hence,

$$N(m, \beta^{-p}) \leq \sum_{(i_1,\dots,i_q) \in \{1,\dots,k\}^q} a_{i_1 i_2} \cdots a_{i_{q-1} i_q} = \sum_{i_1=1}^{k} \sum_{i_q=1}^{k} (A^{q-1})_{i_1 i_q},$$

where $q = p + m - 1$. Using the Jordan form of A, we conclude that there exists a polynomial $c(q)$ such that

$$\sum_{i_1=1}^{k} \sum_{i_q=1}^{k} (A^{q-1})_{i_1 i_q} \leq c(q) \rho(A)^{q-1}.$$

It follows from Theorem 3.5 that

$$h(\sigma|\Sigma_A^+) = \lim_{m \to \infty} \frac{1}{m} \log N(m, \beta^{-p})$$

$$\leq \lim_{m \to \infty} \frac{1}{m} \log[c(q)\rho(A)^{p+m-2}]$$

$$= \log \rho(A).$$

On the other hand, by Proposition 7.3, the number of q-periodic points of $\sigma|\Sigma_A^+$ is equal to $\mathrm{tr}(A^q)$. By (7.8), if ω and ω' are two of these points, then $d_m(\omega, \omega') \geq \beta^{-p}$. Hence,

$$N(m, \beta^{-p}) \geq \mathrm{tr}(A^q)$$

and it follows from Theorem 3.5 that

$$h(\sigma|\Sigma_A^+) = \lim_{m \to \infty} \frac{1}{m} \log N(m, \beta^{-p})$$

$$\geq \lim_{m \to \infty} \frac{1}{m} \log \mathrm{tr}(A^{p+m-1})$$

$$= \lim_{m \to \infty} \frac{1}{m} \log \mathrm{tr}(A^m).$$

Now let $\lambda_1, \ldots, \lambda_k$ be the eigenvalues of A, counted with their multiplicities. Since

$$\mathrm{tr}(A^m) = \sum_{i=1}^{k} \lambda_i^m,$$

we obtain

$$h(\sigma|\Sigma_A^+) \geq \lim_{m \to \infty} \frac{1}{m} \log \sum_{i=1}^{k} \lambda_i^m$$

$$= \log \lim_{m \to \infty} \left(\left| \sum_{i=1}^{k} \lambda_i^m \right|^{1/m} \right)$$

$$= \log \max\{|\lambda_i| : i = 1, \ldots, k\}$$

$$= \log \rho(A).$$

This completes the proof of the theorem.   □

7.3.4 Topological Recurrence

In this section we consider a particular class of topological Markov chains and we study their recurrence properties (see Sect. 3.3). We first introduce two classes of matrices.

Definition 7.5 A $k \times k$ matrix A is called:

1. *irreducible* if for each $i, j \in \{1, \ldots, k\}$ there exists an $m = m(i, j) \in \mathbb{N}$ such that the (i, j)th entry of A^m is positive;
2. *transitive* if there exists an $m \in \mathbb{N}$ such that all entries of the matrix A^m are positive.

Clearly, any transitive matrix is irreducible. However, an irreducible matrix may not be transitive.

Example 7.14 Let

$$A = \begin{pmatrix} 0 & 1 \\ 1 & 0 \end{pmatrix}.$$

No power of A has all entries positive. However, $A^2 = \mathrm{Id}$ and thus, for each pair (i, j), either A or A^2 has positive (i, j)th entry. Hence, the matrix A is irreducible but is not transitive.

Now we consider topological Markov chains with an irreducible or a transitive transition matrix and we study their recurrence properties.

Proposition 7.4 *If the matrix A is irreducible, then the topological Markov chain $\sigma | \Sigma_A^+$ is topologically transitive.*

Proof We first note that the sets

$$D_{j_1 \cdots j_n} = C_{j_1 \cdots j_n} \cap \Sigma_A^+$$
$$= \{\omega \in \Sigma_A^+ : i_m(\omega) = j_m \text{ for } m = 1, \ldots, n\} \tag{7.21}$$

generate the (induced) topology of Σ_A^+. Hence, it is sufficient to consider only these sets in the definition of topological transitivity (see Definition 3.6). Let us then take two nonempty sets $D_{j_1 \cdots j_n}, D_{k_1 \cdots k_n} \subset \Sigma_A^+$. We have to show that there exists an $m \in \mathbb{N}$ such that

$$\sigma^{-m} D_{j_1 \cdots j_n} \cap D_{k_1 \cdots k_n} \neq \varnothing.$$

We first verify that there exists an $m \geq n$ such that the (k_n, j_1)th entry of the matrix A^{m-n+1} is positive (we note that the integer m in Definition 7.5 can be less than n).

Since the matrix A is irreducible, there exist positive integers m_1 and m_2 such that $(A^{m_1})_{k_n j_1} > 0$ and $(A^{m_2})_{j_1 k_n} > 0$. Then

$$
\begin{aligned}
\left(A^{(m_1+m_2)l+m_1}\right)_{k_n j_1} &= \sum_{p=1}^{k} \left(A^{(m_1+m_2)l}\right)_{k_n p} \left(A^{m_1}\right)_{p j_1} \\
&\geq \left(A^{(m_1+m_2)l}\right)_{k_n k_n} \left(A^{m_1}\right)_{k_n j_1} \\
&\geq \left(A^{m_1+m_2}\right)_{k_n k_n}^{l} \left(A^{m_1}\right)_{k_n j_1} \\
&\geq \left(A^{m_1}\right)_{k_n j_1}^{l} \left(A^{m_2}\right)_{j_1 k_n}^{l} \left(A^{m_1}\right)_{k_n j_1} > 0
\end{aligned}
$$

for $l \in \mathbb{N}$ since

$$
\left(A^{m_1+m_2}\right)_{k_n k_n} = \sum_{p=1}^{k} \left(A^{m_1}\right)_{k_n p} \left(A^{m_2}\right)_{p k_n}.
$$

This shows that there exists a transition from k_n to j_1 in

$$
q = (m_1 + m_2)l + m_1
$$

steps. Taking $m = q + n - 1$, we obtain the desired result. Hence, given a sequence $(i_1 i_2 \cdots) \in D_{j_1 \cdots j_n}$, there exist $l_1, \ldots, l_{m-n} \in \{1, \ldots, k\}$ such that

$$
\omega = (k_1 \cdots k_n l_1 \cdots l_{m-n} i_1 i_2 \cdots) \in \Sigma_A^+.
$$

We note that $\omega \in D_{k_1 \cdots k_n}$ and that $\sigma^m(\omega) = (i_1 i_2 \cdots) \in D_{j_1 \cdots j_n}$. Therefore,

$$
\omega \in \sigma^{-m} D_{j_1 \cdots j_n} \cap D_{k_1 \cdots k_n} \neq \varnothing
$$

and the topological Markov chain $\sigma | \Sigma_A^+$ is topologically transitive. $\square$

Now we consider topological Markov chains with a transitive transition matrix.

Proposition 7.5 *If the matrix A is transitive, then the topological Markov chain $\sigma | \Sigma_A^+$ is topologically mixing.*

Proof We proceed in an analogous manner to that in the proof of Proposition 7.4. Given nonempty sets $D_{j_1 \cdots j_n}, D_{k_1 \cdots k_n} \subset \Sigma_A^+$ as in (7.21), we show that there exists a $q \in \mathbb{N}$ such that

$$
\sigma^{-p} D_{j_1 \cdots j_n} \cap D_{k_1 \cdots k_n} \neq \varnothing
$$

for any $p \geq q$.

Lemma 7.1 *If all entries of the matrix A^m are positive, then for each $p \geq m$ all entries of the matrix A^p are positive.*

Proof We first observe that for each $j \in \{1, \ldots, k\}$ there exists an $r = r(j) \in \{1, \ldots, k\}$ such that $a_{rj} = 1$. Otherwise, we would have $(A^p)_{ij} = 0$ for any $p \in \mathbb{N}$ and $i \in \{1, \ldots, k\}$ and thus, the matrix A would not be transitive.

Now we use induction on p. If for some $p \geq m$ the matrix A^p has only positive entries, then

$$\left(A^{p+1}\right)_{ij} = \sum_{l=1}^{k} \left(A^p\right)_{il} a_{lj}$$

$$\geq \left(A^p\right)_{ir} a_{rj} > 0.$$

This completes the proof of the lemma. $\qquad\square$

It follows from Lemma 7.1 that for each $p \in \mathbb{N}$ with $p \geq m + n - 1$, given two nonempty sets $D_{j_1 \cdots j_n}, D_{k_1 \cdots k_n} \subset \Sigma_A^+$, there exist $l_1, \ldots, l_{p-n} \in \{1, \ldots, k\}$ such that

$$\omega = (k_1 \cdots k_n l_1 \cdots l_{p-n} i_1 i_2 \cdots) \in \Sigma_A^+$$

for any sequence $(i_1 i_2 \cdots) \in D_{j_1 \cdots j_n}$. Therefore,

$$\omega \in \sigma^{-p} D_{j_1 \cdots j_n} \cap D_{k_1 \cdots k_n} \neq \varnothing$$

for $p \geq m + n - 1$ and the topological Markov chain $\sigma|\Sigma_A^+$ is topologically mixing. $\qquad\square$

In order to obtain a lower bound for the topological entropy of a topological Markov chain, we need the following simplified version of the Perron–Frobenius theorem.

Theorem 7.2 *Any square matrix with all entries in $\mathbb{N}$ has a real eigenvalue greater than* 1.

Proof Consider the set

$$S = \left\{ v \in (\mathbb{R}_0^+)^k : \|v\| = 1 \right\},$$

where $\|v\| = \sum_{i=1}^{k} |v_i|$ and $v = (v_1, \ldots, v_k)$. Given a $k \times k$ matrix B with all entries b_{ij} in $\mathbb{N}$, we define a function $F \colon S \to S$ by $F(v) = Bv/\|Bv\|$. Since the set S is homeomorphic to the closed unit ball of $\mathbb{R}^{k-1}$ and the function F is continuous, it follows from Brouwer's fixed point theorem[2] that F has a fixed point $v \in S$. Hence,

---

[2]**Theorem** (See for example [26]) Any continuous map $f \colon B \to B$ of a closed ball $B \subset \mathbb{R}^n$ has at least one fixed point.

$Bv = \|Bv\|v$ and v is an eigenvector of B associated to the real eigenvalue

$$\lambda = \|Bv\| = \sum_{i=1}^{k}(Bv)_i$$

$$= \sum_{i=1}^{k}\sum_{j=1}^{k}b_{ij}v_j \geq \sum_{i=1}^{k}\sum_{j=1}^{k}v_j$$

$$= k\sum_{j=1}^{k}v_j = k > 1.$$

This completes the proof of the theorem.   □

One can use Theorem 7.2 to show that any topological Markov chain with a transitive transition matrix has positive topological entropy.

Proposition 7.6 *If the matrix A is transitive, then $h(\sigma|\Sigma_A^+) > 0$.*

Proof Take $m \in \mathbb{N}$ such that A^m has only positive entries. By Theorem 7.2, the matrix A^m has a real eigenvalue $\lambda > 1$. Hence, it follows from Theorem 7.1 that

$$h(\sigma|\Sigma_A^+) = \log\rho(A)$$

$$= \frac{1}{m}\log\rho(A^m)$$

$$\geq \frac{1}{m}\log\lambda > 0.$$

This completes the proof of the proposition.   □

7.4 Horseshoes and Topological Markov Chains

In this section we illustrate how appropriate modifications of the Smale horseshoe give rise to nontrivial topological Markov chains. More precisely, in contrast to all examples of a symbolic coding given in Sect. 7.2, here not all entries of the transition matrix A are equal to 1.

For a better illustration, we consider a specific example. Namely, let f be a diffeomorphism in an open neighborhood of the square $[0,1]^2$ with the behavior shown in Fig. 7.1. We note that one can choose the sizes of the horizontal strips H_i and of their images $V_i = f(H_i)$, for $i = 1, 2, 3$, as well as the diffeomorphism, so that

$$\Lambda = \bigcap_{n\in\mathbb{Z}} f^n(H_1 \cup H_2 \cup H_3)$$

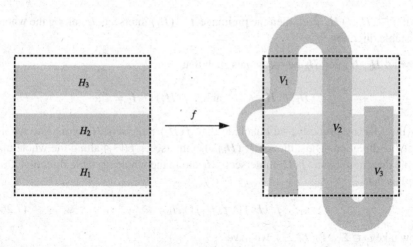

Fig. 7.1 A diffeomorphism f in an open neighborhood of the square $[0, 1]^2$

is a hyperbolic set for f.

Now we consider the 3×3 matrix $A = (a_{ij})$ with entries

$$a_{ij} = \begin{cases} 1 & \text{if } f(H_i) \cap H_j \neq \varnothing, \\ 0 & \text{if } f(H_i) \cap H_j = \varnothing, \end{cases} \tag{7.22}$$

that is,

$$A = \begin{pmatrix} 1 & 0 & 1 \\ 1 & 1 & 1 \\ 1 & 1 & 0 \end{pmatrix}. \tag{7.23}$$

We also consider the set $\Sigma_A \subset \Sigma_3$ induced by this matrix and we define

$$H(\omega) = \bigcap_{n \in \mathbb{Z}} f^{-n} H_{i_n(\omega)}.$$

Proposition 7.7 *The function* $H \colon \Sigma_A \to \Lambda$ *is well defined and*

$$f \circ H = H \circ \sigma \quad \text{in } \Sigma_A. \tag{7.24}$$

Proof Proceeding in an analogous manner to that in Example 7.7, we conclude that

$$\operatorname{card} H(\omega) \leq 1 \quad \text{for } \omega \in \Sigma_A. \tag{7.25}$$

Now we show that card $H(\omega) \geq 1$ for $\omega \in \Sigma_A$. We first note that the following *Markov property* holds:

1. if $f(H_i) \cap H_j \neq \varnothing$, then the image $f(H_i)$ intersects H_j along the whole unstable direction;

2. if $f^{-1}(H_i) \cap H_j \neq \varnothing$, then the preimage $f^{-1}(H_i)$ intersects H_j along the whole stable direction.

Now let H_i, H_j and H_k be rectangles such that

$$f(H_i) \cap H_j \neq \varnothing \quad \text{and} \quad f(H_j) \cap H_k \neq \varnothing.$$

By the Markov property, we conclude that $f(H_i)$ intersects H_j along the whole unstable direction. Thus, the set $f^2(H_i)$ also intersects $f(H_j)$ along the whole unstable direction. Since $f(H_j)$ intersects H_k along the whole unstable direction, this implies that

$$f^2(H_i) \cap f(H_j) \cap H_k \neq \varnothing. \tag{7.26}$$

Now take $\omega \in \Sigma_A$. By (7.22), we have

$$f(H_{i_n(\omega)}) \cap H_{i_{n+1}(\omega)} \neq \varnothing$$

for each $n \in \mathbb{Z}$. By induction, it follows from (7.26) that

$$\bigcap_{k=-n}^{n} f^{n-k}(H_{i_k(\omega)}) \neq \varnothing$$

and

$$K_n := \bigcap_{k=-n}^{n} f^{-k}(H_{i_k(\omega)}) \neq \varnothing.$$

Since the sets K_n are closed and nonempty, the intersection $H(\omega) = \bigcap_{n \in \mathbb{N}} K_n$ is also nonempty and

$$\operatorname{card} H(\omega) = \operatorname{card} \bigcap_{n \in \mathbb{N}} K_n \geq 1.$$

It follows from (7.25) that the function H is well defined.

In order to establish property (7.24), we note that

$$H(\sigma(\omega)) = \bigcap_{n \in \mathbb{Z}} f^{-n}(H_{i_{n+1}(\omega)})$$

$$= \bigcap_{n \in \mathbb{Z}} f^{1-n}(H_{i_n(\omega)})$$

$$= f(H(\omega)).$$

This completes the proof of the proposition.   □

7.5 Zeta Functions

In this section we consider the zeta function of a dynamical system (with discrete time) and in particular of a topological Markov chain.

Definition 7.6 Given a map $f : X \to X$ with

$$a_n := \text{card}\{x \in X : f^n(x) = x\} < \infty \tag{7.27}$$

for each $n \in \mathbb{N}$, its *zeta function* is defined by

$$\zeta(z) = \exp \sum_{n=1}^{\infty} \frac{a_n z^n}{n} \tag{7.28}$$

for each $z \in \mathbb{C}$ such that the series in (7.28) converges.

We recall that the radius of convergence of the power series in (7.28) is given by

$$R = 1 \bigg/ \limsup_{n \to \infty} \sqrt[n]{\frac{a_n}{n}} = 1 \bigg/ \limsup_{n \to \infty} \sqrt[n]{a_n}.$$

In particular, the series converges for $|z| < R$ and the function ζ is holomorphic on the ball $B(0, R) \subset \mathbb{C}$. Clearly, the function ζ is uniquely determined by the sequence $(a_n)_{n \in \mathbb{N}}$ and vice versa.

Now we determine the zeta function of a topological Markov chain.

Example 7.15 Let $\sigma | \Sigma_A^+ : \Sigma_A^+ \to \Sigma_A^+$ be a topological Markov chain defined by a $k \times k$ matrix A with spectral radius $\rho(A) > 0$. It follows from Proposition 7.3 that the sequence $(a_n)_{n \in \mathbb{N}}$ in (7.27) is now $a_n = \text{tr}(A^n)$. Now let $\lambda_1, \ldots, \lambda_k$ be the eigenvalues of A, counted with their multiplicities. We have

$$a_n = \text{tr}(A^n) = \sum_{i=1}^{k} \lambda_i^n.$$

Thus,

$$\zeta(z) = \exp \sum_{i=1}^{k} \sum_{n=1}^{\infty} \frac{\lambda_i^n z^n}{n}$$

$$= \exp \sum_{i=1}^{k} -\log(1 - \lambda_i z)$$

$$= \exp \sum_{i=1}^{k} \log \frac{1}{1 - \lambda_i z}$$

$$= \prod_{i=1}^{k} \frac{1}{1 - \lambda_i z},$$

since

$$\log(1 + w) = \sum_{n=1}^{\infty} \frac{(-1)^n}{n} w^n$$

for $|w| < 1$, where log is the principal branch of the logarithm. On the other hand, the complex numbers $1 - \lambda_i z$ are the eigenvalues of the matrix $\mathrm{Id} - zA$, counted with their multiplicities. This implies that

$$\zeta(z) = \frac{1}{\det(\mathrm{Id} - zA)} \tag{7.29}$$

for

$$|z| < \min\left\{ \frac{1}{|\lambda_i|} : i = 1, \ldots, k \right\} = \frac{1}{\rho(A)}.$$

Example 7.16 The shift map $\sigma : \Sigma_k^+ \to \Sigma_k^+$ coincides with the topological Markov chain defined by the $k \times k$ matrix $A = A_k$ with all entries equal to 1. It follows from (7.29) that

$$\zeta(z) = \frac{1}{\det(\mathrm{Id} - zA_k)}$$

for $|z| < 1/\rho(A_k) = 1/k$. Subtracting the first row of $\mathrm{Id} - zA_k$ from the other rows and then expanding the determinant along the second column, we obtain

$$\det(\mathrm{Id} - zA_k) = \det \begin{pmatrix} 1-z & -z & \cdots & -z \\ -1 & & & \\ \vdots & & \mathrm{Id} & \\ -1 & & & \end{pmatrix}$$

$$= z \det \begin{pmatrix} -1 & 0 & \cdots & 0 \\ -1 & & & \\ \vdots & & \mathrm{Id} & \\ -1 & & & \end{pmatrix} + \det \begin{pmatrix} 1-z & -z & \cdots & -z \\ -1 & & & \\ \vdots & & \mathrm{Id} & \\ -1 & & & \end{pmatrix}$$

$$= -z + \det(\mathrm{Id} - zA_{k-1}).$$

Since $\det(\mathrm{Id} - zA_1) = 1 - z$, it follows by induction that

$$\det(\mathrm{Id} - zA_k) = 1 - kz$$

and thus,

$$\zeta(z) = \frac{1}{1-kz} \quad \text{for } |z| < \frac{1}{k}.$$

Alternatively, the number of n-periodic points of $\sigma | \Sigma_k^+$ is k^n (by Example 7.1) and thus,

$$\zeta(z) = \exp \sum_{n=1}^{\infty} \frac{k^n z^n}{n}.$$

Since

$$\left(\sum_{n=1}^{\infty} \frac{k^n z^n}{n} \right)' = \sum_{n=1}^{\infty} k^n z^{n-1} = \frac{k}{1-kz}$$

for $|z| < 1/k$, we conclude that

$$\zeta(z) = \exp\left[-\log(1-kz) \right] = \frac{1}{1-kz},$$

also for $|z| < 1/k$.

Example 7.17 Now we consider the expanding map $E_2 \colon S^1 \to S^1$. It follows from (2.2) that the number of n-periodic points of E_2 is $2^n - 1$. Hence,

$$\zeta(z) = \exp \sum_{n=1}^{\infty} \frac{(2^n - 1)z^n}{n}.$$

Since

$$\left(\sum_{n=1}^{\infty} \frac{(2^n - 1)z^n}{n} \right)' = \sum_{n=1}^{\infty} (2^n - 1)z^{n-1} = \frac{2}{1-2z} - \frac{1}{1-z}$$

for $|z| < 1/2$, we obtain

$$\zeta(z) = \exp\left[-\log(1 - 2z) + \log(1 - z)\right] = \frac{1 - z}{1 - 2z},$$

also for $|z| < 1/2$.

7.6 Exercises

Exercise 7.1 Determine whether two distances d_β and $d_{\beta'}$ in Σ_k^+ with $\beta \neq \beta'$ can be equivalent.

Exercise 7.2 Show that Σ_A is a closed subset of Σ_k.

Exercise 7.3 Show that (Σ_k^+, d) is a complete metric space.

Exercise 7.4 Determine whether the shift map $\sigma | \Sigma_k$ is topologically mixing.

Exercise 7.5 Let A be a $k \times k$ matrix with entries in $\{0, 1\}$. Show that:

1. if A is irreducible, then $\sigma | \Sigma_A$ is topologically transitive;
2. if A is transitive, then $\sigma | \Sigma_A$ is topologically mixing.

Exercise 7.6 Determine whether the matrix A in (7.23) is irreducible or transitive.

Exercise 7.7 Show that the shift map $\sigma | \Sigma_k^+$ is expansive.

Exercise 7.8 Determine whether the shift map $\sigma | \Sigma_k$ is expansive.

Exercise 7.9 Determine whether:

1. the maps $\sigma | \Sigma_3^+$ and $\sigma | \Sigma_5^+$ are topologically conjugate;
2. the maps $\sigma | \Sigma_k^+$ and $\sigma | \Sigma_k$ are topologically conjugate.

Exercise 7.10 Determine whether the topological Markov chains $\sigma | \Sigma_A^+$ and $\sigma | \Sigma_B^+$ are topologically conjugate for:

1. $A = \begin{pmatrix} 1 & 1 \\ 1 & 0 \end{pmatrix}$ and $B = \begin{pmatrix} 1 & 0 \\ 0 & 1 \end{pmatrix}$;
2. $A = \begin{pmatrix} 1 & 1 \\ 1 & 0 \end{pmatrix}$ and $B = \begin{pmatrix} 0 & 1 \\ 1 & 1 \end{pmatrix}$.

Exercise 7.11 Show that:

1. $h(\sigma | \Sigma_k) = \log k$;
2. the function $H \colon \Sigma_2 \to \Lambda$ defined by (7.18) is a homeomorphism;

3. $h(f|\Lambda) = h(\sigma|\Sigma_2) = \log 2.$

Exercise 7.12 Given an integer $k > 1$, determine whether the set of the topological entropies of all topological Markov chains $\sigma|\Sigma_A^+$ with $\Sigma_A^+ \subset \Sigma_k^+$ contains some interval.

Exercise 7.13 Show that for each integer $m \geq 2$ there exists a continuous map $h: \Sigma_m^+ \to S^1$ such that $h \circ \sigma = E_m \circ h$ in Σ_m^+.

Exercise 7.14 Show that the periodic points of $\sigma|\Sigma_k^+$ are dense.

Exercise 7.15 Show that the periodic points of $\sigma|\Sigma_k$ are dense.

Exercise 7.16 Show that the Smale horseshoe Λ has no isolated points.

Exercise 7.17 Show that the periodic points with even period of the Smale horseshoe are dense.

Exercise 7.18 Construct a symbolic coding for the Smale horseshoe Λ_h in Proposition 5.3.

Exercise 7.19 Compute the zeta function of:

1. the shift map $\sigma|\Sigma_k^+$;
2. the expanding map E_m;
3. the automorphism of the torus $\mathbb{T}^2$ induced by the matrix $\begin{pmatrix} 3 & 1 \\ 2 & 1 \end{pmatrix}$.

Exercise 7.20 Verify that topologically conjugate maps have the same zeta function.

Chapter 8
Ergodic Theory

This chapter gives a first and brief introduction to ergodic theory, avoiding on purpose more advanced topics. After introducing the notions of a measurable map and of an invariant measure, we establish Poincaré's recurrence theorem and Birkhoff's ergodic theorem. We also consider briefly the notions of Lyapunov exponent and of metric entropy. The pre-requisites from measure theory and integration theory are fully recalled in Sect. 8.1.

8.1 Notions from Measure Theory

In this section we recall the necessary notions and results from measure theory and integration theory. Let X be a set and let $\mathcal{A}$ be a family of subsets of X.

Definition 8.1 $\mathcal{A}$ is said to be a σ-*algebra* in X if:

1. $\varnothing, X \in \mathcal{A}$;
2. $X \setminus B \in \mathcal{A}$ when $B \in \mathcal{A}$;
3. $\bigcup_{n=1}^{\infty} B_n \in \mathcal{A}$ when $B_n \in \mathcal{A}$ for each $n \in \mathbb{N}$.

We also consider the σ-algebra *generated* by a family $\mathcal{A}$ of subsets of X: this is the smallest σ-algebra in X containing all elements of $\mathcal{A}$.

Now we introduce the notion of a measure.

Definition 8.2 Given a σ-algebra $\mathcal{A}$ in X, a function $\mu: \mathcal{A} \to [0, +\infty]$ is called a *measure* on X (with respect to $\mathcal{A}$) if:

1. $\mu(\varnothing) = 0$;
2. given pairwise disjoint sets $B_n \in \mathcal{A}$ for $n \in \mathbb{N}$, we have

$$\mu\left(\bigcup_{n=1}^{\infty} B_n\right) = \sum_{n=1}^{\infty} \mu(B_n).$$

L. Barreira, C. Valls, *Dynamical Systems*, Universitext, DOI 10.1007/978-1-4471-4835-7_8, 181
© Springer-Verlag London 2013

We then say that $(X, \mathcal{A}, \mu)$ is a *measure space*.

When the σ-algebra is understood from the context, we still refer to the pair (X, μ) as a measure space.

Example 8.1 Let $\mathcal{A}$ be the σ-algebra in X containing all subsets of X. We define a measure $\mu \colon \mathcal{A} \to \mathbb{N}_0 \cup \{\infty\}$ on X by

$$\mu(B) = \operatorname{card} B.$$

We call μ the *counting measure* on X.

Example 8.2 Now let $\mathcal{B}$ be the *Borel σ-algebra* in $\mathbb{R}$, that is, the σ-algebra generated by the open intervals. Then there exists a unique measure $\lambda \colon \mathcal{B} \to [0, +\infty]$ on $\mathbb{R}$ such that

$$\lambda\big((a, b)\big) = b - a \quad \text{for } a < b.$$

We call λ the *Lebesgue measure* on $\mathbb{R}$.

We also describe a corresponding measure on $\mathbb{R}^n$. Let $\mathcal{B}$ be the Borel σ-algebra in $\mathbb{R}^n$, that is, the σ-algebra generated by the open rectangles $\prod_{i=1}^n (a_i, b_i)$, with $a_i < b_i$ for $i = 1, \ldots, n$. Then there exists a unique measure $\lambda \colon \mathcal{B} \to [0, +\infty]$ on $\mathbb{R}^n$ such that

$$\lambda\left(\prod_{i=1}^n (a_i, b_i) \right) = \prod_{i=1}^n (b_i - a_i)$$

for any $a_i < b_i$ and $i = 1, \ldots, n$. We call λ the *Lebesgue measure* on $\mathbb{R}^n$.

Now we consider measurable functions and their integrals. Let $\mathcal{A}$ be a σ-algebra in the set X.

Definition 8.3 A function $\varphi \colon X \to \mathbb{R}$ is said to be $\mathcal{A}$-*measurable* or simply *measurable* if $\varphi^{-1} B \in \mathcal{A}$ for every $B \in \mathcal{B}$, where $\mathcal{B}$ is the Borel σ-algebra in $\mathbb{R}$.

In order to introduce the notion of the integral of a measurable function, we first consider the class of simple functions. The *characteristic function* of a set $B \subset X$ $\chi_B \colon X \to \{0, 1\}$ is defined by

$$\chi_B(x) = \begin{cases} 1 & \text{if } x \in B, \\ 0 & \text{if } x \notin B. \end{cases}$$

Definition 8.4 Given sets $B_1, \ldots, B_n \in \mathcal{A}$ and numbers $a_1, \ldots, a_n \in \mathbb{R}$, the function

$$s = \sum_{k=1}^n a_k \chi_{B_k}$$

is called a *simple function*.

Clearly, all simple functions are measurable.

Now we introduce the notion of the integral of a nonnegative measurable function.

Definition 8.5 Given a measure space (X, μ), the *(Lebesgue) integral* of a measurable function $\varphi \colon X \to \mathbb{R}_0^+$ is defined by

$$\int_X \varphi \, d\mu = \sup\left\{ \sum_{k=1}^n a_k \mu(B_k) : \sum_{k=1}^n a_k \chi_{B_k} \le \varphi \right\}. \tag{8.1}$$

The integral of an arbitrary function can now be introduced as follows.

Definition 8.6 Given a measure space (X, μ), a measurable function $\varphi \colon X \to \mathbb{R}$ is said to be μ-*integrable* if

$$\int_X \varphi^+ \, d\mu < \infty \quad \text{and} \quad \int_X \varphi^- \, d\mu < \infty,$$

where

$$\varphi^+ = \max\{\varphi, 0\} \quad \text{and} \quad \varphi^- = \max\{-\varphi, 0\}. \tag{8.2}$$

The *(Lebesgue) integral* of a μ-integrable function φ is defined by

$$\int_X \varphi \, d\mu = \int_X \varphi^+ \, d\mu - \int_X \varphi^- \, d\mu. \tag{8.3}$$

8.2 Invariant Measures

In this section we introduce the notion of an invariant measure with respect to a measurable map. We also give several examples of invariant measures.

We first introduce the notion of a measurable map. Let $(X, \mathcal{A}, \mu)$ be a measure space.

Definition 8.7 A map $f \colon X \to X$ is said to be $\mathcal{A}$-*measurable* or simply *measurable* if $f^{-1}B \in \mathcal{A}$ for every $B \in \mathcal{A}$, where

$$f^{-1}B = \{x \in X : f(x) \in B\}.$$

Now we introduce the notion of an invariant measure.

Definition 8.8 Given a measurable map $f \colon X \to X$, we say that μ is f-*invariant* and that f *preserves* μ if

$$\mu\left(f^{-1}B\right) = \mu(B) \quad \text{for } B \in \mathcal{A}. \tag{8.4}$$

Fig. 8.1 A translation of $\mathbb{R}^2$

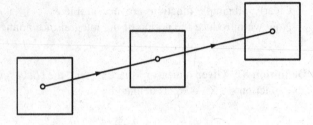

Fig. 8.2 A rotation of $\mathbb{R}^2$

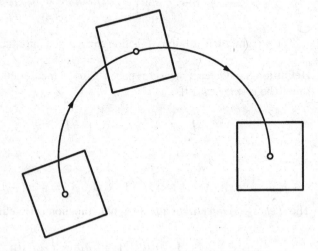

We note that when f is an invertible map with measurable inverse, condition (8.4) is equivalent to

$$\mu(f(B)) = \mu(B) \quad \text{for } B \in \mathcal{A}.$$

Example 8.3 Given $v \in \mathbb{R}^n$, let $f\colon \mathbb{R}^n \to \mathbb{R}^n$ be the translation $f(x) = x + v$ (see Fig. 8.1). Clearly, f is invertible. We also consider the Lebesgue measure λ on $\mathbb{R}^n$. For each $B \in \mathcal{B}$, we have

$$\lambda\big(f(B)\big) = \int_{f(B)} 1\, d\lambda = \int_B |\det d_x f|\, dm(x)$$
$$= \int_B 1\, d\lambda = \lambda(B)$$

and the measure λ is f-invariant. In other words, the translations of $\mathbb{R}^n$ preserve Lebesgue measure.

Example 8.4 Now let $f\colon \mathbb{R}^n \to \mathbb{R}^n$ be a rotation (see Fig. 8.2). Then there exists an $n \times n$ orthogonal matrix A (this means that $A^\mathsf{T} A = \mathrm{Id}$, where A^T is the transpose of A) such that $f(x) = Ax$. Since orthogonal matrices have determinant ± 1, for each $B \in \mathcal{B}$, we have

$$\lambda(f(B)) = \int_{f(B)} 1 \, d\lambda = \int_B |\det d_x f| \, d\lambda(x)$$

$$= \int_B |\det A| \, d\lambda = \int_B 1 \, d\lambda = \lambda(B),$$

where λ is the Lebesgue measure on $\mathbb{R}^n$. Since rotations are invertible maps, this shows that λ is f-invariant and thus, the rotations of $\mathbb{R}^n$ preserve Lebesgue measure.

Example 8.5 Consider a rotation of the circle $R_\alpha : S^1 \to S^1$. Without loss of generality, we assume that $\alpha \in [0, 1]$. We first introduce a measure μ on S^1. For each set $B \subset [0, 1]$ in the Borel σ-algebra in $\mathbb{R}$, we define

$$\mu(B) = \lambda(B). \tag{8.5}$$

Then μ is a measure on S^1 with $\mu(S^1) = 1$. We also have $R_\alpha^{-1} B = B - \alpha$, where

$$B - \alpha = \{x - \alpha : x \in B\} \subset \mathbb{R}.$$

Therefore,

$$\mu(R_\alpha^{-1} B) = \lambda(B - \alpha) = \lambda(B) = \mu(B)$$

since Lebesgue measure is invariant under translations (see Example 8.3). This shows that the rotations of the circle preserve the measure μ.

Example 8.6 For the expanding map $E_m : S^1 \to S^1$, we show that the measure μ introduced in Example 8.5 is E_m-invariant. Given a set $B \subset [0, 1]$ in the Borel σ-algebra in $\mathbb{R}$, we have

$$E_m^{-1} B = \bigcup_{i=1}^{m} B_i, \tag{8.6}$$

where

$$B_i = \left\{ \frac{x+i}{m} : x \in B \right\} \bmod 1$$

(see Fig. 8.3). Since the sets B_i are pairwise disjoint, it follows from (8.6) that

$$\mu(E_m^{-1} B) = \sum_{i=1}^{m} \lambda(B_i)$$

$$= \sum_{i=1}^{m} \frac{\lambda(B+i)}{m} = \sum_{i=1}^{m} \frac{\lambda(B)}{m}$$

$$= \lambda(B) = \mu(B)$$

and the μ measure is E_m-invariant.

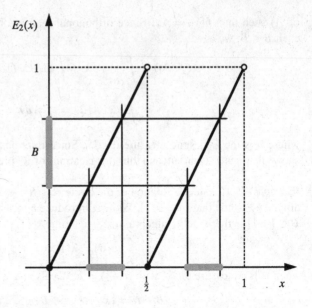

Fig. 8.3 The expanding map E_2 and the preimage $E_2^{-1}B$ of a set B

Example 8.7 The *Gauss map* $f : [0, 1] \to [0, 1]$ is defined by

$$f(x) = \begin{cases} 1/x \bmod 1 & \text{if } x \neq 0, \\ 0 & \text{if } x = 0 \end{cases}$$

(see Fig. 8.4). The map f is closely related to the theory of continued fractions: if

$$x = \cfrac{1}{n_1 + \cfrac{1}{n_2 + \cdots}}$$

is the continued fraction of an irrational number $x \in (0, 1)$, then

$$n_j = \left\lfloor \frac{1}{f^{j-1}(x)} \right\rfloor \quad \text{for } j \in \mathbb{N}.$$

Now we show that the Gauss map preserves the measure μ in $[0, 1]$ defined by

$$\mu(A) = \int_A \frac{1}{1+x} \, dx.$$

We note that it is sufficient to consider the intervals of the form $(0, b)$, with $b \in (0, 1)$ (because they generate the Borel σ-algebra). Since

$$f^{-1}(0, b) = \bigcup_{n=1}^{\infty} \left(\frac{1}{n+b}, \frac{1}{n} \right)$$

Fig. 8.4 The Gauss map

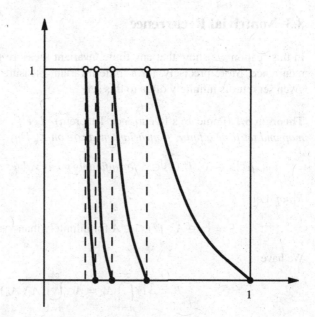

is a disjoint union, we obtain

$$\mu\big(f^{-1}(0,b)\big) = \sum_{n=1}^{\infty} \mu\left(\left(\frac{1}{n+b}, \frac{1}{n}\right)\right)$$

$$= \sum_{n=1}^{\infty} \int_{1/(n+b)}^{1/n} \frac{1}{1+x}\,dx$$

$$= \sum_{n=1}^{\infty} \log \frac{1+1/n}{1+1/(n+b)}$$

$$= \sum_{n=1}^{\infty} \left(\log \frac{n+1}{n+1+b} - \log \frac{n}{n+b}\right)$$

$$= -\log \frac{1}{1+b}$$

$$= \int_{0}^{b} \frac{1}{1+x}\,dx = \mu((0,b))$$

and the measure μ is f-invariant.

8.3 Nontrivial Recurrence

In this section we show that any finite invariant measure gives rise to a nontrivial recurrence. More precisely, for a finite invariant measure almost every point of a given set returns infinitely often to this set.

Theorem 8.1 (Poincaré's Recurrence Theorem) *Let $f\colon X \to X$ be a measurable map and let μ be a finite f-invariant measure on X. For each set $A \in \mathcal{A}$, we have*

$$\mu\big(\{x \in A : f^n(x) \in A \text{ for infinitely many values of } n\}\big) = \mu(A).$$

Proof Let

$$B = \{x \in A : f^n(x) \in A \text{ for infinitely many values of } n\}.$$

We have

$$B = A \cap \bigcap_{n=1}^{\infty} A_n = A \setminus \bigcup_{n=1}^{\infty}(A \setminus A_n), \qquad (8.7)$$

where

$$A_n = \bigcup_{k=n}^{\infty} f^{-k}A.$$

We note that

$$A \setminus A_n \subset A_0 \setminus A_n = A_0 \setminus f^{-n}A_0. \qquad (8.8)$$

Since $A_0 \supset A_n = f^{-n}A_0$ and the measure μ is finite, it follows from (8.8) that

$$0 \le \mu(A \setminus A_n)$$
$$\le \mu\big(A_0 \setminus f^{-n}A_0\big)$$
$$= \mu(A_0) - \mu\big(f^{-n}A_0\big) = 0$$

(because the measure μ is f-invariant). It follows from (8.7) that $\mu(B) = \mu(A)$. $\square$

Now we describe some applications of Theorem 8.1.

Example 8.8 Let $R_\alpha\colon S^1 \to S^1$ be a rotation of the circle. By Example 8.5, the measure μ on S^1 defined by (8.5) is R_α-invariant. Hence, it follows from Theorem 8.1 that, given $c \in [0, 1]$, the set

$$\{x \in [-c, c] : |R_\alpha^n(x)| \le c \text{ for infinitely many values of } n\}$$

has measure $\mu([-c, c]) = 2c$. In other words, almost all points in $[-c, c]$ return infinitely often to $[-c, c]$.

We note that the property established in Example 8.8 is trivial for $\alpha \in \mathbb{Q}$. When $\alpha \in \mathbb{R} \setminus \mathbb{Q}$, it also follows from the density of the orbits of the rotation R_α.

Example 8.9 Consider the expanding map $E_m : S^1 \to S^1$. By Example 8.6, the measure μ on S^1 defined by (8.5) is E_m-invariant. Hence, it follows from Theorem 8.1 that for each interval $[a, b] \subset [0, 1]$, the set

$$\left\{ x \in [a, b] : E_m^n(x) \in [a, b] \text{ for infinitely many values of } n \right\}$$

has measure $\mu([a, b]) = b - a$.

8.4 The Ergodic Theorem

Poincaré's recurrence theorem (Theorem 8.1) says that for a finite invariant measure almost all points of a given set return infinitely often to this set. However, the theorem says nothing about the frequency with which these returns occur. Birkhoff's ergodic theorem establishes the existence of a frequency for almost all points.

Theorem 8.2 (Birkhoff's Ergodic Theorem) *Let $f : X \to X$ be a measurable map and let μ be a finite f-invariant measure on X. Given a μ-integrable function $\varphi : X \to \mathbb{R}$, the limit*

$$\varphi_f(x) \doteq \lim_{n \to \infty} \frac{1}{n} \sum_{k=0}^{n-1} \varphi\left(f^k(x)\right) \tag{8.9}$$

exists for almost every point $x \in X$, the function φ_f is μ-integrable, and

$$\int_X \varphi_f \, d\mu = \int_X \varphi \, d\mu. \tag{8.10}$$

We postpone the proof of Theorem 8.2 until Sect. 8.6.

Example 8.10 Let $f : X \to X$ be a measurable map and let μ be a finite f-invariant measure on X. Given a set $B \in \mathcal{A}$, consider the μ-integrable function $\varphi = \chi_B$. Then

$$\int_X \varphi \, d\mu = \int_X \chi_B \, d\mu = \mu(B)$$

and

$$\varphi_f(x) = \lim_{n \to \infty} \frac{1}{n} \sum_{k=0}^{n-1} \chi_B\left(f^k(x)\right)$$

$$= \lim_{n \to \infty} \frac{1}{n} \operatorname{card}\left\{ k \in \{0, \dots, n-1\} : f^k(x) \in B \right\}.$$

It follows from Theorem 8.2 that

$$\int_X \lim_{n\to\infty} \frac{1}{n} \text{card}\{k \in \{0,\ldots,n-1\}: f^k(x) \in B\}\, d\mu(x) = \mu(B).$$

In this case, the number $\varphi_f(x)$ can be described as the frequency with which the orbit of x visits the set B. Thus, in contrast to Poincaré's recurrence theorem (Theorem 8.1), Birkhoff's ergodic theorem describes in quantitative terms how each orbit returns to the set B.

We also consider briefly the notion of Lyapunov exponent and its relation to Birkhoff's ergodic theorem. Let $f: M \to M$ be a differentiable map.

Definition 8.9 Given $x \in M$ and $v \in T_x M$, the *Lyapunov exponent* of the pair (x, v) is defined by

$$\lambda(x, v) = \limsup_{n\to\infty} \frac{1}{n} \log \|d_x f^n v\|.$$

Now we consider the particular case of the maps of the circle.

Theorem 8.3 *Let $f: S^1 \to S^1$ be a C^1 map and let μ be a finite f-invariant measure on S^1. Then $\lambda(x, v)$ is a limit for almost every x, that is,*

$$\lambda(x, v) = \lim_{n\to\infty} \frac{1}{n} \sum_{k=0}^{n-1} \varphi\big(f^k(x)\big)$$

for almost every $x \in S^1$ and any $v \neq 0$, where $\varphi(x) = \log\|d_x f\|$.

Proof Since the circle S^1 has dimension 1, we have

$$\|d_x f^n v\| = \|d_x f^n\| \cdot \|v\|$$

and thus,

$$\lambda(x, v) = \limsup_{n\to\infty} \frac{1}{n} \log \|d_x f^n\| \tag{8.11}$$

for each $v \neq 0$. Moreover, it follows from the identity

$$d_x f^n = d_{f^{n-1}(x)} f \circ \cdots \circ d_{f(x)} f \circ d_x f$$

that

$$\|d_x f^n\| = \prod_{k=0}^{n-1} \|d_{f^k(x)} f\|.$$

Thus,

$$\frac{1}{n}\log\|d_x f^n\| = \frac{1}{n}\sum_{k=0}^{n-1}\log\|d_{f^k(x)}f\| = \frac{1}{n}\sum_{k=0}^{n-1}\varphi(f^k(x)),$$

where $\varphi(x) = \log\|d_x f\|$. Together with (8.11), this implies that

$$\lambda(x, v) = \limsup_{n\to\infty}\frac{1}{n}\sum_{k=0}^{n-1}\varphi(f^k(x))$$

for each $v \neq 0$. Since φ is continuous, the desired result is now an immediate consequence of Birkhoff's ergodic theorem (Theorem 8.2). □

8.5 Metric Entropy

In this section we consider briefly the notion of metric entropy of an invariant measure. We deliberately present only a (very) simplified version of the theory.

We first introduce the notion of a partition. Let $(X, \mathcal{A}, \mu)$ be a measure space with $\mu(X) = 1$.

Definition 8.10 A finite set $\xi \subset \mathcal{A}$ is called a *partition* of X (with respect to μ) if (see Fig. 8.5):

1. $\mu(\bigcup_{C\in\xi} C) = 1$;
2. $\mu(C \cap D) = 0$ for any $C, D \in \xi$ with $C \neq D$.

Now let $f: X \to X$ be a measurable map preserving the measure μ. Given $n \in \mathbb{N}$ and a partition ξ of X, we construct a new partition ξ_n formed by the sets

$$C_1 \cap f^{-1}C_2 \cap \cdots \cap f^{-(n-1)}C_n,$$

with $C_1, \ldots, C_n \in \xi$.

Definition 8.11 We define

$$h_\mu(f, \xi) = \inf_{n\in\mathbb{N}} -\frac{1}{n}\sum_{C\in\xi_n}\mu(C)\log\mu(C),$$

with the convention that $0\log 0 = 0$.

Example 8.11 Let $f = \mathrm{Id}$. Given $n \in \mathbb{N}$ and a partition ξ of X, we have $\xi_n = \xi$. Thus,

$$h_\mu(f, \xi) = \inf_{n\in\mathbb{N}} -\frac{1}{n}\sum_{C\in\xi}\mu(C)\log\mu(C) = 0.$$

Fig. 8.5 A partition of X

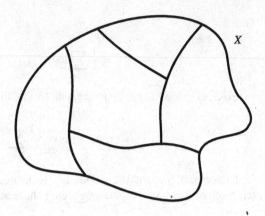

Example 8.12 Consider the expanding map $E_2\colon S^1 \to S^1$ and the E_2-invariant measure μ defined by (8.5). Given $n \in \mathbb{N}$ and the partition

$$\xi = \{[0, 1/2], [1/2, 1]\},$$

we have

$$\xi_n = \left\{\left[\frac{i}{2^n}, \frac{i+1}{2^n}\right] : i = 0, \ldots, 2^n - 1\right\}.$$

Thus,

$$h_\mu(E_2, \xi) = \inf_{n\in\mathbb{N}} -\frac{1}{n} \sum_{C\in\xi_n} \mu(C) \log \mu(C)$$

$$= \inf_{n\in\mathbb{N}} -\frac{1}{n} \cdot 2^n \cdot \frac{1}{2^n} \log \frac{1}{2^n} = \log 2.$$

Finally, we introduce the notion of metric entropy.

Definition 8.12 The *metric entropy* of f with respect to μ is defined by

$$h_\mu(f) = \sup_{n\in\mathbb{N}} h_\mu\big(f, \xi^{(n)}\big),$$

where $\xi^{(n)}$ is any sequence of partitions such that:

1. given $n \in \mathbb{N}$ and $C \in \xi^{(n)}$, there exist $C_1, \ldots, C_m \in \xi^{(n+1)}$ such that

$$\mu\left(C \setminus \bigcup_{i=1}^{m} C_i\right) = \mu\left(\bigcup_{i=1}^{m} C_i \setminus C\right) = 0;$$

2. the union of all partitions $\xi^{(n)}$ generates the σ-algebra $\mathcal{A}$.

One can show that the definition of $h_\mu(f)$ indeed does not depend on the particular sequence $\xi^{(n)}$, but the proof falls outside the scope of this book.

Example 8.13 Let $R_\alpha\colon S^1 \to S^1$ be a rotation of the circle and let μ be the R_α-invariant measure defined by (8.5). Given $n \in \mathbb{N}$ and a partition ξ of X by intervals, we have

$$\operatorname{card}\xi_n \le n \operatorname{card}\xi \tag{8.12}$$

since the endpoints of the intervals in the preimages $f^{-i}\xi$, for $i = 0,\dots,n-1$, determine at most a number $n \operatorname{card}\xi$ of points in S^1. Now we note that

$$-\sum_{C\in\xi_n} \mu(C)\log\mu(C) = \sum_{C\in\xi_n} \varphi(\mu(C)),$$

where

$$\varphi(x) = \begin{cases} -x\log x & \text{if } x \in (0,1], \\ 0 & \text{if } x = 0. \end{cases}$$

Since $\varphi''(x) = -1/x < 0$ for $x \in (0,1)$, the function φ is strictly concave and thus,

$$-\sum_{C\in\xi_n} \mu(C)\log\mu(C) = \sum_{C\in\xi_n} \frac{1}{\operatorname{card}\xi_n}\varphi(\mu(C))\operatorname{card}\xi_n$$

$$\le \varphi\left(\sum_{C\in\xi_n} \frac{\mu(C)}{\operatorname{card}\xi_n}\right)\operatorname{card}\xi_n$$

$$= \varphi\left(\frac{1}{\operatorname{card}\xi_n}\right)\operatorname{card}\xi_n$$

$$= -\log\frac{1}{\operatorname{card}\xi_n} = \log\operatorname{card}\xi_n.$$

Hence, it follows from (8.12) that

$$h_\mu(f,\xi) \le \inf_{n\in\mathbb{N}} \frac{1}{n}\log\operatorname{card}\xi_n$$

$$\le \inf_{n\in\mathbb{N}} \frac{1}{n}\log(n\operatorname{card}\xi) = 0$$

and thus $h_\mu(f) = 0$.

Example 8.14 Consider the expanding map $E_2\colon S^1 \to S^1$ and the E_2-invariant measure μ defined by (8.5). We proceed in an analogous manner to that in Example 8.12. Given $n \in \mathbb{N}$, consider the partition

$$\xi^{(m)} = \left\{\left[\frac{i}{2^m}, \frac{i+1}{2^m}\right] : i = 0,\dots,2^m - 1\right\}.$$

For each $m, n \in \mathbb{N}$, we obtain

$$\xi_n^{(m)} = \left\{ \left[\frac{i}{2^{m+n-1}}, \frac{i+1}{2^{m+n-1}} \right] : i = 0, \ldots, 2^{m+n-1} - 1 \right\} =: \xi^{(m+n-1)}.$$

Thus,

$$h_\mu\big(E_2, \xi^{(m)}\big) = \inf_{n \in \mathbb{N}} -\frac{1}{n} \sum_{C \in \xi_n^{(m)}} \mu(C) \log \mu(C)$$

$$= \inf_{n \in \mathbb{N}} -\frac{1}{n} \sum_{C \in \xi^{(m+n-1)}} \mu(C) \log \mu(C)$$

$$= \inf_{n \in \mathbb{N}} -\frac{1}{n} \cdot 2^{m+n-1} \cdot \frac{1}{2^{m+n-1}} \log \frac{1}{2^{m+n-1}}$$

$$= \inf_{n \in \mathbb{N}} \frac{m+n-1}{n} \log 2 = \log 2.$$

Since the partitions $\xi^{(m)}$ satisfy the hypotheses of Definition 8.12, we conclude that

$$h_\mu(f) = \sup_{m \in \mathbb{N}} h_\mu\big(f, \xi^{(m)}\big) = \log 2.$$

8.6 Proof of the Ergodic Theorem

This section contains a proof of Theorem 8.2. We first establish an auxiliary result.

Lemma 8.1 *Given a μ-integrable function $\psi : X \to \mathbb{R}$, the function $\psi \circ f$ is also μ-integrable and*

$$\int_X (\psi \circ f) \, d\mu = \int_X \psi \, d\mu.$$

Proof Given a set $B \in \mathcal{A}$, condition (8.4) is equivalent to

$$\int_X \chi_{f^{-1}B} \, d\mu = \int_X \chi_B \, d\mu,$$

or equivalently,

$$\int_X (\chi_B \circ f) \, d\mu = \int_X \chi_B \, d\mu \tag{8.13}$$

since $\chi_B \circ f = \chi_{f^{-1}B}$. For a simple function

$$s = \sum_{k=1}^n a_k \chi_{B_k},$$

it follows from (8.13) that

$$\int_X (s \circ f) \, d\mu = \int_X s \, d\mu.$$ (8.14)

Since $\psi = \psi^+ - \psi^-$ and $\psi^+, \psi^- \geq 0$, it follows from (8.3) that it is sufficient to establish the result for nonnegative functions. Let then $\psi : X \to \mathbb{R}_0^+$ be a μ-integrable function. By the definition of the integral in (8.1), there exists a sequence of simple functions $(s_n)_{n \in \mathbb{N}}$ such that:

1. $0 \leq s_n \leq s_{n+1} \leq \psi$ for $n \in \mathbb{N}$, with

$$\lim_{n \to \infty} s_n(x) = \psi(x) \quad \text{for } x \in X;$$

2.

$$\lim_{n \to \infty} \int_X s_n \, d\mu = \int_X \psi \, d\mu.$$ (8.15)

It follows from Fatou's lemma[1] and (8.14) that

$$\int_X \lim_{n \to \infty} (s_n \circ f) \, d\mu \leq \liminf_{n \to \infty} \int_X (s_n \circ f) \, d\mu$$

$$= \liminf_{n \to \infty} \int_X s_n \, d\mu$$

$$= \lim_{n \to \infty} \int_X s_n \, d\mu$$

$$= \int_X \psi \, d\mu < \infty.$$

Hence, the function

$$\lim_{n \to \infty} (s_n \circ f) = \psi \circ f$$

---

[1] **Theorem** (See for example [27]) Given a measure space (X, μ), if $\varphi_n : X \to \mathbb{R}_0^+$ is a sequence of measurable functions, then

$$\int_X \liminf_{n \to \infty} \varphi_n \, d\mu \leq \liminf_{n \to \infty} \int_X \varphi_n \, d\mu.$$

is μ-integrable. Since $s_n \circ f \nearrow \psi \circ f$ when $n \to \infty$, it follows from the monotone convergence theorem[2] that the limit

$$\lim_{n\to\infty} \int_X (s_n \circ f) \, d\mu = \int_X \lim_{n\to\infty} (s_n \circ f) \, d\mu = \int_X (\psi \circ f) \, d\mu \qquad (8.16)$$

exists. Finally, it follows from (8.15) and (8.16) together with (8.14) that

$$\int_X (\psi \circ f) \, d\mu = \lim_{n\to\infty} \int_X (s_n \circ f) \, d\mu$$

$$= \lim_{n\to\infty} \int_X s_n \, d\mu = \int_X \psi \, d\mu.$$

This completes the proof of the lemma. $\square$

Now we consider the set

$$A = \left\{ x \in X : \sup_{n \in \mathbb{N}} \sum_{k=0}^{n-1} \varphi(f^k(x)) > 0 \right\}.$$

Lemma 8.2 *We have* $\int_A \varphi \, d\mu \geq 0$.

Proof The functions $s_0(x) = 0$ and

$$s_n(x) = \sum_{k=0}^{n-1} \varphi(f^k(x)), \qquad \text{for } n \in \mathbb{N},$$

satisfy the identity

$$s_n(f(x)) = s_{n+1}(x) - \varphi(x). \qquad (8.17)$$

Writing

$$t_n(x) = \max\{s_1(x), \dots, s_n(x)\} \quad \text{and} \quad r_n(x) = \max\{0, t_n(x)\},$$

it follows from (8.17) that

$$r_n(f(x)) = t_{n+1}(x) - \varphi(x). \qquad (8.18)$$

On the set

$$A_n = \{x \in X : t_n(x) > 0\},$$

---

[2]**Theorem** (See for example [27]) Given a measure space (X, μ), if $\varphi_n \colon X \to \mathbb{R}_0^+$ is a nondecreasing sequence of measurable functions, then

$$\int_X \lim_{n\to\infty} \varphi_n \, d\mu = \lim_{n\to\infty} \int_X \varphi_n \, d\mu.$$

we have $t_n(x) = r_n(x)$ and thus,

$$\int_{A_n} t_{n+1} \, d\mu \geq \int_{A_n} t_n \, d\mu$$

$$= \int_{A_n} r_n \, d\mu$$

$$= \int_X r_n \, d\mu.$$

We also have

$$\int_{A_n} r_n \, d\mu \leq \int_X r_n \, d\mu.$$

It then follows from (8.18) and Lemma 8.1 that

$$\int_{A_n} \varphi \, d\mu \geq \int_{A_n} t_{n+1} \, d\mu - \int_{A_n} r_n \, d\mu$$

$$\geq \int_X r_n \, d\mu - \int_X (r_n \circ f) \, d\mu = 0. \tag{8.19}$$

Now we note that $A_n \subset A_{n+1}$ for each $n \in \mathbb{N}$ and $\bigcup_{n=1}^{\infty} A_n = A$. Hence, letting $n \to \infty$ in (8.19) we obtain $\int_A \varphi \, d\mu \geq 0$. □

We proceed with the proof of the theorem. Given $a, b \in \mathbb{Q}$ with $a < b$, consider the set

$$B = B_{a,b} = \left\{ x \in X : \liminf_{n \to \infty} \frac{1}{n} \sum_{k=0}^{n-1} \varphi(f^k(x)) < a < b < \limsup_{n \to \infty} \frac{1}{n} \sum_{k=0}^{n-1} \varphi(f^k(x)) \right\}$$

and the function

$$\psi(x) = \begin{cases} \varphi(x) - b & \text{if } x \in B, \\ 0 & \text{if } x \notin B. \end{cases}$$

It follows from Lemma 8.2 that

$$\int_{A_\psi} \psi \, d\mu \geq 0, \tag{8.20}$$

where

$$A_\psi = \left\{ x \in X : \sup_{n \in \mathbb{N}} \frac{1}{n} \sum_{k=0}^{n-1} \psi(f^k(x)) > 0 \right\}$$

$$= \left\{ x \in X : \sup_{n \in \mathbb{N}} \frac{1}{n} \sum_{k=0}^{n-1} \varphi(f^k(x)) > b \right\}.$$

We note that $A_\psi \supset B$. Since $f^{-1}B = B$, we also have

$$\sum_{k=0}^{n-1} \psi\big(f^k(x)\big) = 0 \quad \text{for } x \notin B,$$

that is, $X \setminus B \subset X \setminus A_\psi$. This shows that $A_\psi = B$ and inequality (8.20) is equivalent to

$$\int_B \varphi \, d\mu \geq b\mu(B). \tag{8.21}$$

Analogously, considering the function

$$\bar{\psi}(x) = \begin{cases} a - \varphi(x) & \text{if } x \in B, \\ 0 & \text{if } x \notin B, \end{cases}$$

one can show that

$$\int_B \varphi \, d\mu \leq a\mu(B). \tag{8.22}$$

Since $a < b$, it follows from (8.21) and (8.22) that

$$\mu(B_{a,b}) = \mu(B) = 0.$$

Moreover, since the union of the sets $B_{a,b}$ for $a, b \in \mathbb{Q}$ with $a < b$ coincides with the set of points $x \in X$ such that

$$\liminf_{n \to \infty} \frac{1}{n} \sum_{k=0}^{n-1} \varphi\big(f^k(x)\big) < \limsup_{n \to \infty} \frac{1}{n} \sum_{k=0}^{n-1} \varphi\big(f^k(x)\big),$$

we conclude that the limit $\varphi_f(x)$ in (8.9) exists for almost every $x \in X$.

It remains to establish the integrability of the function φ_f and identity (8.10). Write $\varphi = \varphi^+ - \varphi^-$, with φ^+ and φ^- as in (8.2). Since the functions φ^+ and φ^- are μ-integrable, it follows from the previous argument that the limits

$$\varphi_f^+(x) = \lim_{n \to \infty} \frac{1}{n} \sum_{k=0}^{n-1} \varphi^+\big(f^k(x)\big)$$

and

$$\varphi_f^-(x) = \lim_{n \to \infty} \frac{1}{n} \sum_{k=0}^{n-1} \varphi^-\big(f^k(x)\big)$$

exist for almost every $x \in X$. One can now use Fatou's lemma together with Lemma 8.1 to conclude that

$$\int_X \varphi_f^+ \, d\mu \leq \liminf_{n \to \infty} \frac{1}{n} \sum_{k=0}^{n-1} (\varphi^+ \circ f^k) \, d\mu$$

$$= \liminf_{n \to \infty} \frac{1}{n} \sum_{k=0}^{n-1} \int_X \varphi^+ \, d\mu = \int_X \varphi^+ \, d\mu < \infty$$

and analogously,

$$\int_X \varphi_f^- \, d\mu \leq \int_X \varphi^- \, d\mu < \infty.$$

Thus, the functions φ_f^+ and φ_f^- are μ-integrable and hence, φ_f is also μ-integrable. Finally, we consider the set

$$D_{a,b} = \{x \in X : a \leq \varphi_f(x) \leq b\},$$

for each $a, b \in \mathbb{Q}$ with $a < b$. One can repeat the former argument to show that

$$a\mu(D_{a,b}) \leq \int_{D_{a,b}} \varphi \, d\mu \leq b\mu(D_{a,b}).$$

We also have

$$a\mu(D_{a,b}) \leq \int_{D_{a,b}} \varphi_f \, d\mu \leq b\mu(D_{a,b})$$

and thus,

$$\left| \int_{D_{a,b}} \varphi_f \, d\mu - \int_{D_{a,b}} \varphi \, d\mu \right| \leq (b-a)\mu(D_{a,b}).$$

Hence, given $r > 0$, we obtain

$$\left| \int_X \varphi_f \, d\mu - \int_X \varphi \, d\mu \right| \leq \sum_{n \in \mathbb{Z}} \left| \int_{E_n} \varphi_f \, d\mu - \int_{E_n} \varphi \, d\mu \right|$$

$$\leq \sum_{n \in \mathbb{Z}} r\mu(E_n) = r,$$

where $E_n = D_{nr,(n+1)r}$. Letting $r \to 0$, we conclude that

$$\int_X \varphi_f \, d\mu = \int_X \varphi \, d\mu.$$

8.7 Exercises

Exercise 8.1 For a σ-algebra $\mathcal{A}$, show that if $B_n \in \mathcal{A}$ for $n \in \mathbb{N}$, then $\bigcap_{n=1}^{\infty} B_n \in \mathcal{A}$.

Exercise 8.2 Show that the Borel σ-algebra in $\mathbb{R}$ coincides with the σ-algebra generated by the closed sets in $\mathbb{R}$.

Exercise 8.3 For a measure space $(X, \mathcal{A}, \mu)$, show that:

1. if the sets $B_n \in \mathcal{A}$ satisfy $B_n \subset B_{n+1}$ for $n \in \mathbb{N}$, then

$$\mu\left(\bigcup_{n=1}^{\infty} B_n\right) = \lim_{n \to \infty} \mu(B_n) = \sup_{n \in \mathbb{N}} \mu(B_n);$$

2. if the sets $B_n \in \mathcal{A}$ satisfy $B_n \supset B_{n+1}$ for $n \in \mathbb{N}$ and $\mu(B_1) < \infty$, then

$$\mu\left(\bigcap_{n=1}^{\infty} B_n\right) = \lim_{n \to \infty} \mu(B_n) = \inf_{n \in \mathbb{N}} \mu(B_n).$$

Exercise 8.4 Given a set X and a point $p \in X$, let

$$\delta_p(B) = \begin{cases} 1 & \text{if } p \in B, \\ 0 & \text{if } p \notin B \end{cases}$$

for each $B \subset X$. Show that:

1. δ_p is a measure on the σ-algebra of all subsets of X;
2. any function $\varphi\colon X \to \mathbb{R}$ is measurable;
3. any function $\varphi\colon X \to \mathbb{R}$ is δ_p-integrable and

$$\int_X \varphi \, d\delta_p = \varphi(p).$$

Exercise 8.5 Verify that any translation of a set in the Borel σ-algebra $\mathcal{B}$ in $\mathbb{R}$ is still in $\mathcal{B}$.

Exercise 8.6 Show that:

1. any continuous function $\varphi\colon \mathbb{R} \to \mathbb{R}$ is $\mathcal{B}$-measurable;
2. any monotonous function $\varphi\colon \mathbb{R} \to \mathbb{R}$ is $\mathcal{B}$-measurable.

Exercise 8.7 Show that a function $\varphi\colon \mathbb{R} \to \mathbb{R}$ is $\mathcal{B}$-measurable if and only if

$$\{x \in \mathbb{R} : \varphi(x) > \alpha\} \in \mathcal{B} \quad \text{for } \alpha \in \mathbb{R}.$$

Exercise 8.8 Show that the sum and the product of measurable functions are still measurable functions. Hint: Use Exercise 8.7.

Exercise 8.9 Show that the supremum and the limit of a sequence $\varphi_n \colon \mathbb{R} \to \mathbb{R}$ of measurable functions are still measurable functions. Hint: Note that

$$\left\{ x \in \mathbb{R} : \sup_{n \in \mathbb{N}} \varphi_n(x) \leq \alpha \right\} = \bigcap_{n \in \mathbb{N}} \{ x \in \mathbb{R} : \varphi_n(x) \leq \alpha \}$$

and

$$\left\{ x \in \mathbb{R} : \lim_{n \to \infty} \varphi_n(x) \leq \alpha \right\} = \bigcap_{k \in \mathbb{N}} \bigcup_{n \in \mathbb{N}} \bigcap_{m \geq n} \{ x \in \mathbb{R} : \varphi_m(x) \leq \alpha + 2^{-k} \}.$$

Exercise 8.10 Show that a measurable function φ is integrable if and only if $|\varphi|$ is integrable.

Exercise 8.11 Show that if the function $\varphi \colon X \to \mathbb{R}$ is μ-integrable, then

$$\left| \int_X \varphi \, d\mu \right| \leq \int_X |\varphi| \, d\mu.$$

Exercise 8.12 Show that a point $x \in [0, 1]$ is rational if and only if $f^m(x) = 0$ for some $m \in \mathbb{N}$, where f is the Gauss map.

Exercise 8.13 Let $f \colon \mathbb{R}^n \to \mathbb{R}^n$ be a C^1 diffeomorphism. Show that f preserves Lebesgue measure if and only if $|\det d_x f| = 1$ for every $x \in \mathbb{R}^n$.

Exercise 8.14 Verify that Poincaré's recurrence theorem cannot be generalized to infinite measure spaces.

Exercise 8.15 Let $f \colon X \to X$ be a measurable map and let μ be a finite f-invariant measure on X. Show that if the function $\varphi \colon X \to \mathbb{R}$ is μ-integrable, then

$$\lim_{n \to \infty} \frac{\varphi(f^n(x))}{n} = 0$$

for almost every $x \in X$.

Exercise 8.16 Let $f \colon X \to X$ be a measurable map preserving a measure μ on X with $\mu(X) = 1$. Show that if ξ is a partition of X, then $h_\mu(f, \xi) \leq \log \operatorname{card} \xi$.

Exercise 8.17 Compute the metric entropy of the expanding map $E_m \colon S^1 \to S^1$ with respect to the E_m-invariant measure μ defined by (8.5).

Exercise 8.18 Show that any automorphism of the torus $\mathbb{T}^n$ preserves the measure induced on $\mathbb{T}^n$ by the Lebesgue measure λ on $\mathbb{R}^n$.

Exercise 8.19 Show that any endomorphism of the torus $\mathbb{T}^n$ preserves the measure induced on $\mathbb{T}^n$ by the Lebesgue measure λ on $\mathbb{R}^n$.

Exercise 8.20 Given constants $p_1, \ldots, p_k > 0$ with $\sum_{i=1}^{k} p_i = 1$, consider the measure μ on Σ_k^+ defined by

$$\mu(C_{i_1 \cdots i_n}) = p_{i_1} \cdots p_{i_n}$$

for each set $C_{i_1 \cdots i_n}$ in (7.5). Show that:

1. μ is σ-invariant and $\mu(\Sigma_k^+) = 1$;
2. $h_\mu(\sigma) = -\sum_{i=1}^{k} p_i \log p_i$.

References

1. Abraham, R., Marsden, J.: Foundations of Mechanics. Benjamin/Cummings, Redwood City (1978)
2. Alsedà, L., Llibre, J., Misiurewicz, M.: Combinatorial Dynamics and Entropy in Dimension One. Advanced Series in Nonlinear Dynamics, vol. 5. World Scientific, Singapore (2000)
3. Amann, H.: Ordinary Differential Equations: An Introduction to Nonlinear Analysis. de Gruyter Studies in Mathematics, vol. 13. de Gruyter, Berlin (1990)
4. Anderson, J.: Hyperbolic Geometry. Springer Undergraduate Mathematics Series. Springer, London (2005)
5. Arnol'd, V.: Geometrical Methods in the Theory of Ordinary Differential Equations. Grundlehren der Mathematischen Wissenschaften, vol. 250. Springer, New York (1988)
6. Arnol'd, V.: Mathematical Methods of Classical Mechanics. Graduate Texts in Mathematics, vol. 60. Springer, New York (1989)
7. Barreira, L.: Dimension and Recurrence in Hyperbolic Dynamics. Progress in Mathematics, vol. 272. Birkhäuser, Basel (2008)
8. Barreira, L.: Thermodynamic Formalism and Applications to Dimension Theory. Progress in Mathematics, vol. 294. Birkhäuser, Basel (2011)
9. Barreira, L.: Ergodic Theory, Hyperbolic Dynamics and Dimension Theory. Universitext. Springer, Berlin (2012)
10. Barreira, L., Pesin, Ya.: Lyapunov Exponents and Smooth Ergodic Theory. University Lecture Series, vol. 23. Am. Math. Soc., Providence (2002)
11. Barreira, L., Pesin, Ya.: Nonuniform Hyperbolicity: Dynamics of Systems with Nonzero Lyapunov Exponents. Encyclopedia of Mathematics and Its Applications, vol. 115. Cambridge University Press, Cambridge (2007)
12. Barreira, L., Valls, C.: Ordinary Differential Equations: Qualitative Theory. Graduate Studies in Mathematics, vol. 137. Am. Math. Soc., Providence (2012)
13. Beardon, A.: The Geometry of Discrete Groups. Graduate Texts in Mathematics, vol. 91. Springer, New York (1983)
14. Beardon, A.: Iteration of Rational Functions: Complex Analytic Dynamical Systems. Graduate Texts in Mathematics, vol. 132. Springer, Berlin (1991)
15. Bonatti, C., Díaz, L., Viana, M.: Dynamics Beyond Uniform Hyperbolicity: A Global Geometric and Probabilistic Perspective. Encyclopaedia of Mathematical Sciences, vol. 102. Springer, Berlin (2005)
16. Bowen, R.: Equilibrium States and Ergodic Theory of Anosov Diffeomorphisms. Lecture Notes in Mathematics, vol. 470. Springer, Berlin (1975)
17. Brin, M., Stuck, G.: Introduction to Dynamical Systems. Cambridge University Press, Cambridge (2002)

18. Brown, J.: Ergodic Theory and Topological Dynamics. Pure and Applied Mathematics, vol. 70. Academic Press, New York (1976)
19. Carleson, L., Gamelin, T.: Complex Dynamics. Universitext. Springer, New York (1993)
20. Chernov, N., Markarian, R.: Chaotic Billiards. Mathematical Surveys and Monographs, vol. 127. Am. Math. Soc., Providence (2006)
21. Chicone, C.: Ordinary Differential Equations with Applications. Texts in Applied Mathematics, vol. 34. Springer, New York (1999)
22. Chow, S.-N., Hale, J.: Methods of Bifurcation Theory. Grundlehren der Mathematischen Wissenschaften, vol. 251. Springer, New York (1982)
23. Coddington, E., Levinson, N.: Theory of Ordinary Differential Equations. McGraw-Hill, New York (1955)
24. Cornfeld, I., Fomin, S., Sinai, Ya.: Ergodic Theory. Grundlehren der Mathematischen Wissenschaften, vol. 245. Springer, Berlin (1982)
25. de Melo, W., van Strien, S.: One-Dimensional Dynamics. Ergebnisse der Mathematik und ihrer Grenzgebiete, vol. 25. Springer, Berlin (1993)
26. Dieck, T.: Algebraic Topology. Textbooks in Mathematics. Eur. Math. Soc., Zürich (2008)
27. Friedman, A.: Foundations of Modern Analysis. Dover, New York (1982)
28. Guckenheimer, J., Holmes, P.: Nonlinear Oscillations, Dynamical Systems, and Bifurcations of Vector Fields. Applied Mathematical Sciences, vol. 42. Springer, New York (1983)
29. Hale, J., Magalhães, L., Oliva, W.: Dynamics in Infinite Dimensions. Applied Mathematical Sciences, vol. 47. Springer, New York (2002)
30. Hale, J., Verduyn Lunel, S.: Introduction to Functional-Differential Equations. Applied Mathematical Sciences, vol. 99. Springer, New York (1993)
31. Irwin, M.: Smooth Dynamical Systems. Pure and Applied Mathematics, vol. 94. Academic Press, New York (1980)
32. Katok, A., Hasselblatt, B.: Introduction to the Modern Theory of Dynamical Systems. Encyclopedia of Mathematics and Its Applications, vol. 54. Cambridge University Press, Cambridge (1995)
33. Katok, A., Strelcyn, J.-M., with the collaboration of Ledrappier, F., Przytycki, F.: Invariant Manifolds, Entropy and Billiards: Smooth Maps with Singularities. Lecture Notes in Mathematics, vol. 1222. Springer, Berlin (1986)
34. Katok, S.: Fuchsian Groups. Chicago Lectures in Mathematics. University of Chicago Press, Chicago (1992)
35. Keller, G.: Equilibrium States in Ergodic Theory. London Mathematical Society Student Texts, vol. 42. Cambridge University Press, Cambridge (1998)
36. Kitchens, B.: Symbolic Dynamics: One-Sided, Two-Sided and Countable State Markov Shifts. Universitext. Springer, Berlin (1998)
37. Lind, D., Marcus, B.: An Introduction to Symbolic Dynamics and Coding. Cambridge University Press, Cambridge (1995)
38. Mañé, R.: Ergodic Theory and Differentiable Dynamics. Ergebnisse der Mathematik und ihrer Grenzgebiete, vol. 8. Springer, Berlin (1987)
39. McMullen, C.: Complex Dynamics and Renormalization. Annals of Mathematics Studies, vol. 135. Princeton University Press, Princeton (1994)
40. Milnor, J.: Dynamics in One Complex Variable. Annals of Mathematics Studies, vol. 160. Princeton University Press, Princeton (2006)
41. Morosawa, S., Nishimura, Y., Taniguchi, M., Ueda, T.: Holomorphic Dynamics. Cambridge Studies in Advanced Mathematics, vol. 66. Cambridge University Press, Cambridge (2000)
42. Moser, J.: Stable and Random Motions in Dynamical Systems. Annals of Mathematics Studies, vol. 77. Princeton University Press, Princeton (1973)
43. Munkres, J.: Topology: A First Course. Prentice-Hall, New York (1975)
44. Nitecki, Z.: Differentiable Dynamics: An Introduction to the Orbit Structure of Diffeomorphisms. MIT Press, Cambridge (1971)
45. Palis, J., de Melo, W.: Geometric Theory of Dynamical Systems: An Introduction. Springer, New York (1982)

46. Palis, J., Takens, F.: Hyperbolicity and Sensitive Chaotic Dynamics at Homoclinic Bifurcations: Fractal Dimensions and Infinitely Many Attractors. Cambridge Studies in Advanced Mathematics, vol. 35. Cambridge University Press, Cambridge (1993)
47. Parry, W.: Topics in Ergodic Theory. Cambridge University Press, Cambridge (1981)
48. Parry, W., Pollicott, M.: Zeta Functions and the Periodic Orbit Structure of Hyperbolic Dynamics. Astérisque, vol. 187–188. Soc. Math. France, Montrouge (1990)
49. Pesin, Ya.: Dimension Theory in Dynamical Systems: Contemporary Views and Applications. Chicago Lectures in Mathematics. Chicago University Press, Chicago (1997)
50. Pesin, Ya.: Lectures on Partial Hyperbolicity and Stable Ergodicity. Zürich Lectures in Advanced Mathematics. Eur. Math. Soc., Zürich (2004)
51. Petersen, K.: Ergodic Theory. Cambridge Studies in Advanced Mathematics, vol. 2. Cambridge University Press, Cambridge (1983)
52. Pollicott, M.: Lectures on Ergodic Theory and Pesin Theory on Compact Manifolds. London Mathematical Society Lecture Note Series, vol. 180. Cambridge University Press, Cambridge (1993)
53. Pollicott, M., Yuri, M.: Dynamical Systems and Ergodic Theory. London Mathematical Society Student Texts, vol. 40. Cambridge University Press, Cambridge (1998)
54. Przytycki, F., Urbanski, M.: Conformal Fractals: Ergodic Theory Methods. London Mathematical Society Lecture Note Series, vol. 371. Cambridge University Press, Cambridge (2010)
55. Robinson, C.: Dynamical Systems: Stability, Symbolic Dynamics, and Chaos. CRC Press, Boca Raton (1994)
56. Rudolph, D.: Fundamentals of Measurable Synamics: Ergodic Theory on Lebesgue Spaces. Oxford Science Publications. Oxford University Press, Oxford (1990)
57. Ruelle, D.: Thermodynamic Formalism. Encyclopedia of Mathematics and Its Applications, vol. 5. Addison-Wesley, Reading (1978)
58. Schmidt, K.: Dynamical Systems of Algebraic Origin. Progress in Mathematics, vol. 128. Birkhäuser, Basel (1995)
59. Sell, G., You, Y.: Dynamics of Evolutionary Equations. Applied Mathematical Sciences, vol. 143. Springer, New York (2002)
60. Shub, M.: Global Stability of Dynamical Systems. Springer, New York (1986)
61. Sinai, Ya.: Introduction to Ergodic Theory. Mathematical Notes, vol. 18. Princeton University Press, Princeton (1976)
62. Sinai, Ya.: Topics in Ergodic Theory. Princeton Mathematical Series, vol. 44. Princeton University Press, Princeton (1994)
63. Szlenk, W.: An Introduction to the Theory of Smooth Dynamical Systems. Wiley, New York (1984)
64. Temam, R.: Infinite-Dimensional Dynamical Systems in Mechanics and Physics. Applied Mathematical Sciences, vol. 68. Springer, New York (1997)
65. Walters, P.: An Introduction to Ergodic Theory. Graduate Texts in Mathematics, vol. 79. Springer, Berlin (1982)

Index